ACS SYMPOSIUM SERIES **454**

Interactions of Food Proteins

Nicholas Parris, EDITOR
Robert Barford, EDITOR
U.S. Department of Agriculture

Developed from a symposium sponsored
by the 1989 International Chemical Congress
of Pacific Basin Societies,
Honolulu, Hawaii,
December 17–22, 1989

American Chemical Society, Washington, DC 1991

CHEM
Sep / ae

Library of Congress Cataloging-in-Publication Data

Interactions of food proteins
Nicholas Parris, editor, Robert Barford, editor.

p.   cm.—(ACS symposium series; 454)
"Developed from a symposium sponsored by the 1989 International Chemical Congress of Pacific Basin Societies, Honolulu, Hawaii, December 17–22, 1989."

Includes bibliographical references and index.

ISBN 0–8412–1935–4 : $59.95
1. Proteins.   I. Parris, Nicholas, 1937– . II. Barford, Robert A., 1936– . III. International Chemical Congress of Pacific Basin Societies (1989 : Honolulu, Hawaii) IV. Series

TP453.P7I55   1991
664 .092—dc20   90–24457
                                         CIP

The paper used in this publication meets the minimum requirements of American National Standard for Information Sciences—Permanence of Paper for Printed Library Materials, ANSI Z39.48–1984.   ∞

SD 2/6/92 RL

Foreword

THE ACS SYMPOSIUM SERIES was founded in 1974 to provide a medium for publishing symposia quickly in book form. The format of the Series parallels that of the continuing ADVANCES IN CHEMISTRY SERIES except that, in order to save time, the papers are not typeset, but are reproduced as they are submitted by the authors in camera-ready form. Papers are reviewed under the supervision of the editors with the assistance of the Advisory Board and are selected to maintain the integrity of the symposia. Both reviews and reports of research are acceptable, because symposia may embrace both types of presentation. However, verbatim reproductions of previously published papers are not accepted.

Contents

Preface ix

1. Relationship of Composition to Protein Functionality 1
   Karen L. Fligner and Michael E. Mangino

2. Significance of Macromolecular Interaction and Stability
   in Functional Properties of Food Proteins 13
   Akio Kato

3. Effect of Preheat Temperature on the Hydrophobic
   Properties of Milk Proteins 25
   N. Parris, J. H. Woychik, and P. Cooke

4. Quantitation of Hydrophobicity for Elucidating the
   Structure–Activity Relationships of Food Proteins 42
   S. Nakai, E. Li-Chan, M. Hirotsuka, M. C. Vazquez,
   and G. Arteaga

5. Milk Protein Ingredients: Their Role in Food Systems 59
   Steve J. Haylock and Wayne B. Sanderson

6. Significance of Lysozyme in Heat-Induced Aggregation
   of Egg White Protein 73
   Naotoshi Matsudomi

7. Formation and Interaction of Plant Protein Micelles
   in Food Systems 91
   M. A. H. Ismond, S. D. Arntfield, and E. D. Murray

8. Diffusion and Energy Barrier Controlled Adsorption
   of Proteins at the Air–Water Interface 104
   Srinivasan Damodaran and Kyung B. Song

9. Surface Activity of Bovine Whey Proteins
   at the Phospholipid–Water Interface 122
   Donald G. Cornell

10. Interactions Between Milk Proteins and Lipids:
    A Mobility Study.. 137
    M. Le Meste, B. Closs, J. L. Courthaudon, and
    B. Colas

11. Some Aspects of Casein Micelle Structure...................................... 148
    Lawrence K. Creamer

12. Cross-Linkage Between Casein and Colloidal Calcium
    Phosphate in Bovine Casein Micelles... 164
    Takayoshi Aoki

13. Quaternary Structural Changes of Bovine Casein
    by Small-Angle X-ray Scattering: Effect of Genetic Variation 182
    T. F. Kumosinski, H. Pessen, E. M. Brown,
    L. T. Kakalis, and H. M. Farrell, Jr.

14. Genetic Engineering of Bovine κ-Casein To Enhance
    Proteolysis by Chymosin 195
    Sangsuk Oh and Tom Richardson

15. Rheology: A Tool for Understanding Thermally Induced
    Protein Gelation 212
    D. D. Hamann

16. Food Dough Constant Stress Rheometry 228
    Jimbay Loh

17. Factors Influencing Heat-Induced Gelation of Muscle
    Proteins.. 243
    Denise M. Smith

18. Gelation of Myofibrillar Protein 257
    E. Allen Foegeding, Clark J. Brekke, and
    Youling L. Xiong

19. Interactions of Muscle and Nonmuscle Proteins Affecting
    Heat-Set Gel Rheology 268
    Tyre C. Lanier

Author Index 285

Affiliation Index 285

Subject Index.. 286

Preface

BECAUSE OF THE CONTINUING TREND toward new and improved food formulation and the need for food ingredients that have a wide range of functional properties, this volume was written with two goals in mind: to describe the physicochemical changes that take place in processed foods and to elucidate the interactions that occur at a molecular level.

A manufactured food product is a complex colloidal system in which structural and textural properties are determined by the number and strength of interactions between various types of macromolecules within the system. It is therefore important to examine adsorption and aggregation of macromolecules at the colloidal level, protein denaturation, the interfacial behavior of proteins and smaller molecules with lipids, and the rheological properties of the final product.

This book is intended for food scientists in industry, academia, and government. An understanding of physicochemical changes and how such changes affect food performance will allow development of treatments to optimize food performance.

The nineteen chapters of this book discuss the effect of processing conditions on macromolecular interactions, colloidal stability, and functional and rheological properties in various food systems. When selecting speakers for this symposium, we attempted to include experts from both industry and academia. We offer our apologies to those experts who were missed for various reasons. We also thank the authors for their excellent presentations and their timely preparation of the chapters.

NICHOLAS PARRIS
ROBERT BARFORD
Eastern Regional Research Center
Agricultural Research Service
U.S. Department of Agriculture
Philadelphia, PA 19118

November 2, 1990

Chapter 1

Relationship of Composition to Protein Functionality

Karen L. Fligner and Michael E. Mangino

Department of Food Science and Technology, Ohio State University, Columbus, OH 43210

Proteins are responsible for many of the desirable attributes associated with food products. Performance results from the interactions of proteins with other proteins or food components. Food proteins are commonly composed of several fractions, often having different properties. Thus, a particular functional property may be a manifestation of a specific component of the food protein used. Functionality has been defined as "any property of a food or food ingredient, except its nutritional ones, that affects its utilization" (1). The domain of physical functions associated with the presence of proteins in a food system typically includes: (a) increased hydration and water binding which affect viscosity and gelation, (b) modification of surface tension and interfacial activity which control emulsification and foaming ability and (c) chemical reactivity leading to altered states of cohesion/adhesion and a potential for texturization. Functional properties are influenced by intrinsic factors, such as composition, conformation and homogeneity of the protein source, as well as by processing methods and environmental conditions (2). There are several specific papers that relate protein structure to function (1-5). Reports that relate composition to function are rare. This chapter will discuss the effects of compositional factors of proteins, with emphasis on whey proteins, on gelation, foaming and emulsification. Composition encompasses certain physicochemical properties such as, hydrophobicity and sulfhydryl content, as well as specific ratios of protein components, ratios of protein to other constituents, such as lipid, carbohydrate and mineral components and the presence of other components in the food product including added emulsifiers.

Gelation

Protein gelation is a two stage-process involving an initial denaturation or unfolding of protein molecules followed by subsequent aggregation. Gel formation can occur when protein-protein interactions lead to the formation of a three-dimensional network capable of entraining water molecules. A balance between the attractive forces necessary to form a network and the repulsive forces necessary to prevent its collapse is

0097–6156/91/0454–0001$06.00/0

required for gel formation ($\underline{6,7}$). The major compositional factors important to protein gel formation include calcium content, free sulfhydryl content and protein hydrophobicity. The ratios of carbohydrate, nonprotein nitrogen and lipid to protein content and the amount and type of whey protein utilized also influence gelation ($\underline{8}$, $\underline{9}$).

Several workers ($\underline{9}$-$\underline{12}$) have reported that addition of electrolytes to a protein system, such as calcium and sodium salts, can affect the strength and texture of whey protein concentrate (WPC) gels. At low calcium concentrations, weak gels were formed, while calcium concentrations up to 11 mM increased gel strength. Schmidt and Morris ($\underline{9}$), Modler ($\underline{13}$), Modler and Jones ($\underline{14}$), and Schmidt et al. ($\underline{15}$) found that the addition of 5-20 mM $CaCl_2$ or 0.1 to 0.3M NaCl increased the gel strength of heated WPC. The addition of greater than 25 mM $CaCl_2$ or 0.4 M NaCl caused a decrease in gel strength. It has been proposed that calcium provides ionic bridges within the gel matrix that increase gel strength. The presence of excess calcium may cause protein aggregation before protein unfolding can occur and prevents the formation of a three-dimensional network ($\underline{11}$). Two reviews ($\underline{13,15}$) describe the replacement of calcium with sodium to yield WPC with increased solubility and increased gelation time at 70°C. As calcium replacement increased, gelation time increased. In addition, replacement of calcium with sodium improved the textural properties of the gels. DeWit et al. ($\underline{16}$) also demonstrated a relationship between ionic strength and gel formation. Removal of calcium and lactose before thermal processing followed by addition of calcium salts prior to gelation resulted in improved gel formation by whey proteins. An inverse relationship between the extent of protein denaturation and the effectiveness of gelation was also observed.

Langley and Green ($\underline{17}$) examined the effect of protein content and composition on the compression and strength of whey protein gels. The composition of powders varied from 0-12% α-lactalbumin (α-La), 44-87% β-lactoglobulin (β-Lg) and 6-56% casein derived proteins. Compressive strength, elastic modulus and impact strength increased with increasing β-Lg content. The free sulfhydryl and disulfide bond content of β-Lg may account for this as these bonds can break and reform with heat denaturation ($\underline{17}$). The sulfhydryl content of WPC has been related to the strength and textural characteristics of WPC gels ($\underline{17,18}$). The addition of sulfhydryl blocking or modifying agents resulted in decreased gel strength, suggesting that sulfhydryls are important to gelation ($\underline{9}$). Zirbel and Kinsella ($\underline{19}$) noted that the addition of thiol reagents altered hardness, elasticity and cohesiveness of whey protein gels, supporting previous observations that disulfide bonds are an important compositional factor. As was the case with calcium, an optimum concentration for maximum gel strength has been reported ($\underline{10}$). Schmidt and Morris ($\underline{9}$) found that the effect of disulfide reducing agents is concentration dependent. Low concentrations enhanced gelation, while higher concentrations impaired gel formation. In addition, the appearance of whey protein gels has been related to total sulfhydryl content. Low concentrations resulted in clear, translucent gels, while higher concentrations yielded opaque gels. Storage of WPC at room temperature resulted in decreased total sulfhydryls. However, Kohnhorst and Mangino ($\underline{20}$) showed that the variability in the concentration of sulfhydryl groups found in commercially available WPC did not correlate to the gel strength of these samples.

Hashizume and Sato ($\underline{21,22}$) showed that the gel forming properties of milk proteins were related to the number of accessible sulfhydryls. Heating skim milk below 70°C caused no change in accessible sulfhydryl groups, while heating at 80°C or greater increased sulfhydryl accessibility. Gel firmness was affected by changes in accessibility of sulfhydryl groups. Gel irmness was highest between 2×10^{-6} and 4×10^{-6} mol SH groups/g protein. They concluded that disulfide bond formation was due to intermolecular sulfhydryl-disulfide interchange and not due to the oxidation of free sulfhydryls to form new disulfide bonds. Mangino et al. ($\underline{23}$) found that pasteurization of the rententate resulted in decreased native β-Lg which correlated well with observed decreases in free sulfhydryl content of the resulting WPC. Gel strength at pH 8.0 was also decreased.

The effects of sulfhydryl groups on WPC gel strength has been shown to be related to pH (16, 24). At low pH values, there is relatively little effect of sulfhydryl content on gel strength, while at pH values near the pK of the sulfhydryl group an effect has been noted. Schmidt and coworkers (9) have also described the pH dependence of WPC gel appearance and texture in detail. At low pH, an opaque coagulum forms while at high pH a translucent, viscoelastic gel results. Shimada and Cheftel (25) proposed that hydrophobic interactions and intermolecular disulfide bonds are responsible for the resulting network of whey protein isolate (WPI) gels at neutral and alkaline pH's. At low pH, hydrogen bonds are believed to be significant. Zirbel and Kinsella (19) reported that the addition of ethanol to whey protein gels altered gel hardness, suggesting hydrogen bonding and electrostatic interactions were important. In addition, gel texture was related to free sulfhydryl content; however, a pH dependence was observed. Gel firmness decreased with decreasing sulfhydryl content at pH values ranging from 6.5 to 9.5, while elasticity increased.

Mulvihill and Kinsella (11) observed that lipids and lactose impair gelation. Gel characteristics of whey proteins prepared by ultrafiltration were dependent on salt content and pH. Removal of lactose and salts improved gelation at pH 6.0. However, increased removal impaired gelation at pH 8.0. Schmidt (18) noted that differences in composition were not entirely responsible for differences in gelling time in WPC produced by various methods. Preparation technique, in relation to denaturation level, was also important. The protein/lactose and protein/fat ratio might account for differences in gelation characteristics of WPC produced by gel filtration and elecrodialysis. Stronger, more translucent, cohesive and springy gels were produced from WPC which had lower levels of lactose and ash. DeWit et al. (16) observed that lactose decreased gel strength. However, this could be overcome by increased temperature for gelation.

Protein crosslinks can also be formed through hydrophobic interactions. Similar to the effects noted with calcium and sulfhydryl content, there appears to be an optimal level of hydrophobicity beyond which gel strength is weakened (20, 24, 26, 27). In most WPC's, this optimal value has not been reached and increases in protein hydrophobicity result in increased gel strength (24). Joseph and Mangino (28, 29) have demonstrated that the concentration of milk fat globule membrane (MFGM) derived proteins in WPC can be directly correlated with gel strength, presumably due to their contribution to protein hydrophobicity. In model systems containing β-Lg and calcium, the addition of membrane proteins increased gel strength until a maximal value was reached. Further addition caused a decrease in gel strength. Commercial WPC have values of hydrophobicity below the optimal range and factors that increase their hydrophobicity also increase gel strength.

The above data suggests that the observed gelation behavior of whey proteins is in reasonable agreement with the theoretical assessment of the forces involved in gel formation. In many cases, the effects observed from a treatment or a physical or chemical manipulation of the structure of a protein or its environment are dependent on the extent of the treatment. Both positive and negative effects on gel strength can be observed. It is also apparent that in complex systems interactions of components make an important contribution to the final characteristics of the system. Data obtained from relatively simple one or two component model systems may not agree with data obtained from more complex food systems. Consideration should always be given to the possibility that the presence of other gross components may influence the effect of specific additives or treatments on the gel characteristics of food proteins. Further study on the mechanism of food protein gelation and on factors that affect the texture of the resulting gels is warranted. The important contributions that proteins make to the structure and texture of fabricated foods through the formation of gels may well make the gelation properties of food proteins one of their potentially most valuable attributes.

Emulsification

To fully understand the function of proteins as emulsifying agents, it is important to distinguish between the measured parameters for emulsion stability (ES), emulsification activity index (EAI), emulsifying activity (EA) and emulsifying capacity (EC). The following descriptions will prove helpful. ES can be defined as the maintenance of a homogeneous structure and texture of the system. Several techniques are available to determine if an emulsion has undergone any observable changes and commonly involve, but are not limited to, measurement of the amount of oil or cream separation from an emulsion during a given time period at a particular temperature and gravitational field. EAI is expressed as the area of interface (determined from turbidity measurements) stabilized per unit weight of protein. EA can be defined as the total surface area of the emulsion. EC is generally defined as the maximum amount of lipid emulsified by a protein dispersion. Oil is added at a given rate to a protein dispersion until the viscosity decreases or inversion occurs.

Emulsions are thermodynamically unstable mixtures of immiscible substances. When lipid and water are mixed there is a strong driving force to limit contact between them and phase separation occurs. Minimal contact can initially be achieved by the formation of spherical droplets through the input of work. Small droplet formation is facilitated by the addition of small molecular weight emulsifiers which reduce interfacial tension. Proteins stabilize emulsions through the ordered structuring of water molecules which results in minimum contact of hydrophobic groups with water; the energetically most favorable state. For a thorough discussion of the role of proteins in the stabilization of emulsions, see Mangino (30). Adsorption of proteins at an interface is believed to occur in three stages. The native protein first diffuses to the contact region where it penetrates the interface and surface denaturation results. The adsorbed protein rearranges to form the lowest free energy state by inserting hydrophobic groups into the oil phase (31, 32). This results in formation of a film of protein around the fat globules.

Important variables for protein absorption at interfaces include conformational flexibility, hydrophobicity and viscoelasticity of the interfacial film. Changes in pH and ionic strength can affect hydrophobic properties through alteration of protein conformation. Kato and Nakai (33) first reported a correlation between surface hydrophobicity and emulsifying activity. Others have confirmed this observation (34-36). Voutsinas et al. (37) found that high surface hydrophobicity and high solubility were important for optimum emulsifying properties as measured by EAI. Marshall and Harper (38) concluded that partial unfolding of proteins to expose hydrophobic groups resulted in improved emulsifying properties. Lee and Kim (39) investigated several methods for determining hydrophobicity and related them to surface activity. A relationship between emulsifying index and surface hydrophobicity as determined by the cis-parinaric acid technique was observed. Partition and hydrophobic chromatography also correlated with surface activity but the ANS probe technique did not. In general, proteins that possess high levels of surface hydrophobicity exhibit favorable emulsification characteristics.

The hydrophobicity of WPC has been related to the content of MFGM proteins present in the WPC. Joseph and Mangino (29) have reported a positive correlation between MFGM content in WPC and effective hydrophobicity. Marshall and Harper (38) confirm the contribution of MFGM proteins to WPC hydrophobicity. Kanno (40) examined the emulsifying properties of bovine MFGM in milk fat emulsions. In reconstituted MFG stabilized with MFGM, it was observed that as MFGM increased from 20 to 80 mg MFGM/g fat, ES and EA increased, while EAI decreased. Maximum EA and ES occurred at 2% MFGM (80 mg MFGM/g fat) and at pH 4.0.

These reports indicate that relatively small amounts of membrane protein in the WPC can have an important effect on the hydrophobicity of the WPC and positively contribute to their ability to form emulsions in food systems.

The emulsifying properties of whey proteins and various WPC were related to the compositional differences resulting from different processing techniques by de Wit et al. (16). They observed that α-La was a satisfactory emulsifier even though its surface hydrophobicity was low. A propensity for surface denaturation may account for this. BSA and immunoglobulin G (IgG) were poor emulsifiers due to their large molecular weights. Increased denatured protein level, as long as solubility was not greatly affected, correlated with high EAI. Thus, unfolding exposed hydrophobic groups to enhance emulsification.

Shimizu and co-workers (41, 42), Yamauchi et al. (43) and Das and Kinsella (44) studied the pH dependent emulsifying properties of whey proteins. Shimizu observed that the surface hydrophobicity of β–Lg was greater at pH 3.0 than at pH 7.0. However, the adsorption rate at pH 3.0 was slower. EAI and surface hydrophobicity were negatively correlated at low pH. Cleavage of disulfide bonds improved emulsification properties at pH 3.0. An increase in average fat globule diameter at lower pH values was also observed. Selective adsorption, which may reflect effective hydrophobiciy, occurred at all pH values studied. In addition, pH changes may induce conformational changes that can affect hydrophobicity (43). A correlation between adsorbability and average hydrophobicity of individual whey proteins as calculated by the Bigelow method was not observed. For α–La, surface hydrophobicity and adsorbability increased as the pH range was decreased from 9.0 to 3.0. However, for other proteins, surface hydrophobicity did not correlate with adsorbability. High adsorption rates were found at low pH values for α-La and high pH values for β-Lg. Das and Kinsella (44) obtained similar results.

The explanation for the above findings may be related to known conformational characteristics of this protein. At lower pH values, the structure of β-Lg is rigid, while in the alkaline region, the structure is more flexible enabling favorable hydrophobic interactions with the interface even though the solution hydrophobicity is low (44). Shimizu and co-authors have proposed that average hydrophobicity is not meaningful for studying the adsorption of proteins due to the pH dependence, whereas conformational flexibility is more informative. Kinsella and Whitehead (45) and Dickinson and Stainsby (46) also concluded that the emulsifying behavior of β-Lg correlated better with molecular flexibility than with hydrophobicity. It was suggested that surface hydrophobicity governs emulsifying activity, while flexibility is important for foaming. Thus, emulsification is related to the accessibility of hydrophobic groups at the surface rather than to the total number of hydrophobic groups present. In practical terms, Kinsella and Whitehead (45) have drawn attention to the important differences in the methods for evaluating foams and emulsions and the possible effects the evaluation method can have on the conclusions drawn.

A descriptive, general review on emulsion and foam stability has been presented by Halling (47). A major point relates protein film rigidity to stability of emulsions against coalescence. Proteins containing disulfide links, such as κ-casein and β-Lg, adsorb more slowly and resist complete unfolding. Thus, there is more room at the interface to accommodate larger amounts of protein. Waniska and Kinsella (48) demonstrated that maximum surface viscosity of β-Lg solutions occurred at the isoelectric point (pI). At the pI, there were few intermolecular interactions enabling hydrophobic residues to stabilize a more compact tertiary structure. Near the pI, proteins form close-packed, condensed films. Thus, increased protein concentration at the interface resulting from ionic and hydrophobic interactions will potentially cause the formation of a more viscous interfacial film which is believed to be beneficial for emulsion stability.

Differing abilities of proteins to stabilize emulsions may be due to different mechanisms of adsorption. For example, Tornberg, as reported by Toro-Vazquez and Regenstein (49), has demonstrated that soy proteins form an associated complex while caseinates migrate to the interface as monomers. She claims that isolated caseinates are less aggregated than casein in NFDM, even though both possess the same overall composition in terms of protein fractions. In contrast, Morr (50) theorizes that caseins and caseinates function in an associated form and not as monomers. Thus, Morr (50) has concluded that fat globules are likely stabilized through adsorption of intact subunits. It follows that factors that favor dissociation of highly aggregated forms of the protein increase functionality.

Toro-Vazquez and Regenstein (49) found that protein solubility and surface hydrophobicity were important in maximizing EAI at low ionic strength, while at high ionic strength rigidity or disulfide bond content was more important. Similar results were observed when ES was the dependent variable. High chloride ion content is believed to affect hydrophobic interactions by (a) chaotropically altering the structure of water and exposing more hydrophobic groups to the aqueous phase and (b) decreasing electrostatic interactions that are unfavorable for protein adsorption. Matsudomi et al. (51) have confirmed that changes in hydrophobicity occur with temperature and ionic strength. Hydrophobicity is greater at low ionic strength than at high ionic strength for soy glycinin. They suggest that at low ionic strength, proteins possessing low disulfide-bond content should stabilize emulsions by charge repulsion because the proteins should lie flat on the surface, while at high ionic strength these proteins should adsorb with more loops and tails. Proteins with high disulfide-bond content unfold less and exhibit higher viscoelasticity. A balance of free and bound sulfhydryls may be important for adequate viscoelasticity. Marshall and Harper (38) suggested that reduction or cleavage of disulfide bonds can increase functionality by the formation of aggregates through recombining of SH groups. Morr (50) suggested that disulfide-interchange reactions decrease functionality by decreasing flexibility.

Dickinson et al. (52) found that β-Lg is more difficult to displace from interfaces than α-La due to greater surface viscosity. The presence of free sulfhydryl groups on each subunit of β-Lg is believed to contribute to the greater surface viscosity of β-Lg. Once β-Lg is unfolded at the interface, disulfide bonds can link together subunits to result in a structured film making it difficult to displace. Dickinson (53) reported that a strong viscoelastic film is required for stability. He rated lysozyme>κ-casein>β-casein in order of effectiveness in preventing coalescence. This represents a positive correlation between viscoelasticity of the protein film and stability towards coalescence. Halling (47) also reported that κ-casein>α_{s1}-casein>β-casein in terms of creaming stability. Graham and Phillips (31) found that in terms of emulsion stability, globular proteins produce more stable emulsions in the order of BSA>lysozyme>β-casein. Leman et al. (54) showed that emulsions based on β-Lg and WPI were more resistant to creaming than systems based on skim milk proteins (SMP) or caseins. The least satisfactory cream stability was observed in emulsions stabilized by micellar casein. Furthermore, β-Lg and WPI also required less energy input to produce satisfactory emulsion stability. In addition, Leman et al. (54) determined a decreasing order of cream stability of β-Lg>WPI>SMP>casein micelles. In this connection, it must be noted that ionic strength is important since the addition of 15 mM NaCl increased creaming stability and decreased again at 25 mM.

Whether or not proteins act alone or in concert with other functional additives has been addressed in a number of reports. Fligner and Mangino (55) evaluated the relationship of phospholipid, carrageenan and casein-to-whey protein ratio on the physical stability of concentrated infant formula. Protein type was more important than phospholipid or carrageenan in controlling stability. Specifically, higher whey protein content produced more stable emulsions, provided the level of whey protein did not exceed the concentration leading to gel formation upon thermal processing. Other studies by Fligner and Mangino (56), have shown that as lecithin level is increased from 0.2% to 0.5% stability increases. However, as additional lecithin was incorporated up to 0.8%, stability decreased. Yamauchi et al. (43) also observed that addition of surfactants in whey protein emulsions, including lecithin at 0.005%, decreased stability at pH 7.0. They concluded that lecithin interfered with adsorption of β-Lg. The enhancement of emulsifying properties of various globular proteins by formation of a protein/egg-yolk/lecithin complex were investigated by Nakamura et al. (57). Emulsifiying properties were enhanced at pH 3.0 for all proteins studied except ovomucoid and lysozyme. They postulate that a rigid structure, due to the presence of disulfide bonds, may limit the formation of the complex. Therefore, hydrophobic groups could not be exposed. Addition of reducing agents improved emulsifying properties. Thus, lysozyme hydrophobicity may increase with reducing agent treatment.

Dickinson and Stainsby (58) have recognized that food systems containing both protein and polysaccharide may exhibit thermodynamic incompatibility. Tolstuguzov (59, 60) has reviewed this area thoroughly and has suggested that molecules prefer to be surrounded by like molecules and when the concentration is high enough, phase separation may occur. In spite of such considerations, it remains that long term stability depends on the thickness and strength of the interfacial film. In addition, film aging is believed to favor stability because film strength increases over time, possibly through the formation of crosslinks (58).

Protein modifications can be related to factors suspected to be important to functionality, such as alterations in charge density and hydrophobicity. Work in this area holds promise for clarifying the importance of compositional factors and providing information on the molecular forces involved. Jimenz-Flores and Richardson (61) discussed genetic engineering of caseins as a means to alter functionality. For example, the content of phosphate esters in β-casein may be genetically altered to improve its emulsifying ability. Dickinson and Stainsby (58), Arai and Watanabe (62), Mitchell (63) and Kinsella and Whitehead (45) reviewed chemical and enzymatic modifications of proteins and their effects on functionality. A large positive correlation was found between the capacity of esterified proteins to decrease interfacial tension and hydrophobicity as determined by the ANS-probe technique. The methyl derivative was most effective suggesting enhanced surface hydrophobicity improved the hydrophobic/hydrophilic balance.

It should be recognized that alterations in protein structure may lead to drastic changes in conformation, which in turn may produce unexpected results. For example, increased hydrophilicity in some cases, can play a role in film formation. Specifically, glycosylation of β-Lg increased molecular weight, decreased charge and surface hydrophobicity resulting in improved emulsifying properties.

Disulfide bonds provide structural constraints to unfolding. Therefore, alteration of disulfide bond content may affect functional properties. Reduction of disulfide bonds can increase surface hydrophobicity through molecular rearrangement. However, it has been shown that disulfide bridges contribute to structured regions that result in more cohesive films and improve long term stability. Lysozyme, for example, is a globular, rigid protein with a low surface hydrophobicity contrasted with the more surface active β-casein which is disordered, flexible and possesses a high surface hydrophobicity. Lysozyme forms concentrated films, whereas β-casein forms dilute films (31, 32).

The nature of emulsion instability can be quite different in various food products. The practical implications of applying the emulsifying ability of proteins in food formuation are considerable and must be viewed in the context of their effects on emulsion formation and subsequent stability. In some cases, the emulsion may totally fail with resultant phase separation. In other cases, instability is manifested by a clustering of fat globules that results in cream layer formation due to density differences. In some products, this may only be a minor problem, while for others, it may be the factor that limits shelf-life. Thus, any study that attempts to define factors or conditions that are important to emulsion formation or stability must carefully define the system under study and the type of instability that is being observed.

The added complexity of the system when proteins are present often makes a clear understanding of the relative importance of various components and processes to emulsion formation and stabilization difficult to interpret. Much progress has been made in relating structure to function and several contributing factors have been established. In some cases, protein type has proven to be more important for emulsion stability than traditional emulsifying agents. The mechanism of emulsification stabilization by proteins involves (a) adsorption at the interface to provide a film of high flexibility and viscoelasticity and (b) an aging factor to permit favorable cross-linking. The hydrophobic/hydrophilic balance of the protein is of paramount importance in the selection of the food protein ingredient. However, the protein must be chosen with due consideration to other ingredients present, such as surfactants and hydrophilic colloids.

Foaming

Foam formation is similar in many ways to emulsification. In order for foam formation to occur, water molecules must surround air droplets. When water molecules are forced into contact with relatively nonpolar air, they tend to become more ordered resulting in a high surface tension and a high surface energy. Proteins decrease this surface energy by interacting with both air and water molecules. Foams initially contain a large number of small air cells separated by an absorbed interfacial layer and by areas of water. As the foam ages, water drains and the air cells approach each other. However, the presence of an interfacial layer will continue to provide stability even after considerable drainage has occurred.

During whipping, hydrophobic regions of proteins interact with air and cause unfolding and exposure of buried hydrophobic groups. This decreases surface tension which allows for creation of more bubbles. The partially unfolded proteins then associate to form a film around the air droplets (64). Compositional factors important in foaming include surface hydrophobicity and the spatial distribution of hydrophobicity, lipid, mineral, carbohydrate and sulfhydryl content (65). According to Poole and Fry (64), the ideal foaming protein possesses high surface hydrophobicity, good solubility and a low net charge at the pH of the food product. It must also be easily surface denatured. In general, stable foams occur near the isoelectric point of the protein where electrostatic repulsion is minimal (66). Thus, compositional factors that minimize protein aggregation and increase solubility in the isoelectric range (for example decreased calcium levels) result in improved foaming properties (15).

Phillips et al. (67) examined the effects of milk proteins on the foaming properties of egg white to identify foam depressant components present in whey and milk powders. Previous work (68) suggested that the proteose-peptone component in these powders served as a foam depressant. The components of the proteose-peptone did not decrease overrun of egg white foams. In fact, the proteose-peptone fraction obtained from milk by acid hydrolysis increased overrun. A foam depressant, tentatively identified as a heat-stable, proteolysis-resistant lipoprotein, with a molecular weight of > 160,000 daltons, was isolated from milk, whey and casein. This fraction could be removed by employing ultrafiltration with a 100,000 molecular weight cut-off membrane. Removal of this fraction resulted in improved foaming properties. In addition, β-Lg, sodium caseinate, BSA, lactoferrin and xanthine oxidase increased the overrun of egg white after 15 minutes of whipping. Joseph and Mangino (28, 29) demonstrated a strong inverse correlation between the MFGM protein content of WPC and their foaming properties. It was also noted that the addition of membrane proteins to solutions of egg white greatly inhibited foam formation.

Other antifoaming agents present in milk proteins include unesterified free fatty acids, such as mono- and diglycerides (69) and lipid and phospholipid from the fat globule membrane (13). Defatted WPC exhibit desirable whipping and foaming properties with no oxidized flavor from phospholipids. Modler (13) and Modler and Jones (14) discussed phospholipid removal of ultrafiltered WPC by complexing with CMC. Mangino et al. (23) found that WPC prepared from pasteurized milk contain decreased levels of neutral lipid compared to WPC prepared from raw milk. The reduction in lipid resulted in increased overrun and foam stability.

Lorient et al. (70) examined the foaming properties of individual caseins with respect to pH, ionic strength and polarity. The structural properties of modified caseins were related to the functional behavior of caseinates. They ranked β-casein>α_s-casein>κ-casein in increasing order of effectiveness of foam capacity and formation. Only κ-casein formed thick, stable foams, which was attributed to the presence of thiol groups that enhanced film strength. In addition, chemical glycosylation, increased voluminosity and viscosity to allow formation of thick films.

Poole (71) studied the effect of a mixture of basic and acidic proteins on foaming. Such mixtures allowed the incorporation of up to 30% lipid without detriment to either overrun or foam stability. To be effective, the basic protein should have a pI of 9 or greater and a molecular weight of at least 4,000 daltons. Furthermore, chitosan, a basic polysaccharide, enhanced foaming properties as effectively as the basic proteins, clupeine and lysozyme. The ratio of acidic to basic proteins was important and depended on the pI of the protein utilized. Clupeine (pI=12) was superior to lysozyme (pI=10.7). Good foams were attainable with 4% WPI, 65% sucrose, 0.4% clupeine and containing from 10 and 35% lipid. The foaming properties of chemically modified β-Lg with pI's ranging from 5.3 to 9.9 were also studied. At a pI of 9.9, modified β-Lg was as effective as clupeine. Sucrose and chitosan demonstrated synergistic effects. Phillips et al. (72), Poole et al. (73), Clark et al. (66) and Mitchell (63) discussed similar observations with respect to basic proteins and foaming.

Processes that increase hydrophobicity, such as partial denaturation, enhance foaming properties. The energy barrier to adsorption is decreased when an increasing number of nonpolar residues are exposed at the surface (15, 63). In addition, heat denaturation can improve flexibility (63). Partial denaturation at 40 to 65°C for 30 minutes improved functionality of whey proteins by increasing surface hydrophobicity according to de Wit, as reported by Poole and Fry (64). Similar results were obtained for ovalbumin where surface hydrophobicity increased from 20 to 2200. However, BSA (Ho=1450) was still superior, suggesting that other factors play a critical role. Voutsinas et al., as reported by Mitchell (63), and Kinsella and Whitehead (45) found a curvilinear relationship between surface hydrophobicity and foam stability. This suggests that heat denaturation enhances foam capacity by enhancing the surface activity of proteins possessing low surface hydrophobicities.

Poole and Fry (64) have described chemical and enzymatic modifications that improve foaming properties by charge alteration and exposure of buried hydrophobic groups, respectively. Covalent attachment of a hydrophobic group (chain length of C4-C8) has been shown to result in improved foaming ability suggesting an important role for increased hydrophobicity (62, 63). Bacon et al. (65) related foaming properties to structural differences between various lysozymes and α-La. Increased surface charge was detrimental to foaming. Positive correlations were observed between foaming properties of proteins and surface hydrophobicity. However, it was found that bovine α-La produced better foams than lysozyme even though it possesses poor surface hydrophobicity. Reduction of disulfides in bovine α–La was thought to be of minor significance.

Townsend and Nakai, as reported by Mitchell (63), demonstrated that decreased molecular flexibility (increased disulfide bonds) resulted in increased foam capacity. Foam capacity was inversely related to molecular flexibility in terms of the number of disulfide bonds per unit molecular weight (63). Schmidt et al. (15) has confirmed that reducing agents decrease foamability of WPC, while oxidizing agents increase it. Hansen and Black (74) have demonstrated the existence of an optimal level of oxidation beyond which foaming properties deteriorate. In whey protein/ carboxymethylcellulose complexes, addition of 0.1% H_2O_2 resulted in maximum whipping characteristics. However, addition of 0.2% H_2O_2 caused a dramatic decrease in foam development. Thus, flexibility and total hydrophobicity (instead of surface hydrophobicity) have been proposed as major contributing factors for desirable foaming characteristics of a protein (45, 63).

Conclusions

The behavior of proteins at air/water interfaces has been studied extensively. The forces involved are well characterized and there is considerable knowledge regarding changes that occur at the molecular level. However, a large percentage of this work has been performed with well-defined systems containing very dilute protein solutions. The translation of this information to the more complex situation that exists in foods has also begun. Of considerable importance to the utilization of proteins in food products is the stability of the foam once formed. Often times, foams can be easily made; however, they are extremely unstable. Thus, a substantial effort has been made to understand the mechanism of foam formation and decay and to attempt to define means of extending foam life. In many cases, foams are stabilized by surface denaturation of the protein. In other instances, heat is applied to denature the protein and to stabilize the foam structure. Therefore, in applications, such as the substitution of whey proteins for egg white in the formation of angel food cakes, the protein must not only entrap enough air to obtain the desired overrun, but must also be heat denatured before foam collapse occurs. The area of foam stabilization is receiving considerable attention and it is essential that progress be made in this area before wide-spread protein substitutions will be possible.

Acknowledgments

Salaries and research support provided by state and federal funds appropriated to The Ohio Agricultural Research and Development Center, The Ohio State University. Supported in part by The National Dairy Promotion and Research Board. Journal Article Number 43-90.

Literature Cited

1. Pour-El, A., Ed. Functionality and Protein Structure; American Chemical Society: Washington, DC, 1979; p ix.
2. Kinsella, J. E. In Food Proteins; Fox, P.; Condon J., Eds.; Applied Science Publishers: New York, 1982; pp 51-103.
3. Leman, J.; Kinsella J. CRC Crit. Rev. Food Sci. Nutr. 1989, 28(2), 115-138.
4. Kinsella, J. E. CRC Crit. Rev. Food Sci. Nutr. 1984, 21(3), 197-262.

5. Kinsella, J. E. Crit. Rev. Food Sci. Nutr. 1976, 7(3), 219-280.
6. Zirbel, F.; Kinsella, J. Milchwissenschaft 1988, 43(11), 691-694.
7. Mangino, M. E. J. Dairy Sci. 1984, 67(11), 2711-2722.
8. Morr, C. V. In Developments in Dairy Chemistry-1; Fox, P., Ed.; Applied Science Publishers: New York, 1982; pp. 375-399.
9. Schmidt, R. H.; Morris, H. Food Tech. 1984 5, 85-96.
10. Schmidt, R. H.; Illingworth, B.; Deng, J.; Cornell, J. J. Agric. Food Chem. 1979, 27(3), 529-532.
11. Mulvihill, D. M.; Kinsella, J. Food Tech. 1987, 41(9), 102-111.
12. Johns, J. E. M.; Ennis, B. New Zealand J. Dairy Sci. Tech. 1981, 16, 79-86.
13. Modler, H. W. J. Dairy Sci. 1985, 68(9), 2206-2214.
14. Modler, H. W.; Jones, J. Food Tech. 1987, 41(10), 114-117, 129.
15. Schmidt, R. H.; Packard, V.; Morris, H. J. Dairy Sci. 1984, 67(11), 2723-2733.
16. de Wit, J. N.; Hontelez-Backx, E.; Adamse, M. Neth. Milk Dairy J. 1988, 42, 155-172.
17. Langley, K. R.; Green, M. J. Dairy Res. 1989, 56, 275-284.
18. Schmidt, R. H. In Protein Functionality in Foods; Cherry, J., Ed.; American Chemical Society: Washington, DC, 1981; pp. 131-147.
19. Zirbel, F.; Kinsella, J. Food Hydrocolloids 1988, 2(6), 467-475.
20. Kohnhorst, A. L.; Mangino, M. J. Food Sci. 1985, 50, 1403-1405.
21. Hashizume, K.; Sato, T. J. Dairy Sci. 1988, 71(6), 1439-1446.
22. Hashizume, K.; Sato, T. J. Dairy Sci. 1988, 71(6), 1447-1454.
23. Mangino, M. E.; Liao, Y.; Harper, N.; Morr, C.; Zadow, J. J. Food Sci. 1987, 52(6), 1522-1524.
24. Mangino, M. E.; Kim, J.; Dunkerley, J.; Zadow, J. Food Hydrocolloids 1987, 1(4), 277-282.
25. Shimada, K.; Cheftel, J. J. Agric. Food Chem. 1988, 36, 1018-1025.
26. Voutsinas, L. P.; Nakai, S.; Harwalkar, V. Can. Inst. Food Sci. Tech. J. 1983, 16(3), 185-190.
27. Mangino, M. E.; Fritsch, D.; Liao, S.; Fayerman, A.; Harper, W. New Zealand J. Dairy Sci. Tech. 1985, 20, 103-107.
28. Joseph, M. S. B.; Mangino, M. Austr. J. Dairy Tech. 1988, 5, 9-11.
29. Joseph, M. S. B.; Mangino, M. Austr. J. Dairy Tech. 1988, 5, 6-8.
30. Mangino, M. E. In Food Proteins: Structure and Functional Relationships; Kinsella, J.; Soucie, W., Eds.; Amer. Oil Chem. Soc.: Champaign, Illinois, 1990.
31. Graham, D. E.; Phillips, M. In Theory and Practice of Emulsion Technology; Academic Press: New York, 1976; pp. 75-98.
32. Phillips, M. C. Food Tech. 1981, 1, 50-57.
33. Kato, A.; Nakai, S. Biochim. Biophys. Acta 1980, 624, 13-20.
34. Kato, A.; Osako, Y.; Matsudomi, N.; Kobayashi, K. Agric. Biol. Chem. 1983, 47, 33-37.
35. Nakai, S. J. Agric. Food Chem. 1983, 31, 676-683.
36. Li-Chan, E.; Nakai, S; Wood, D. J. Food Sci. 1984, 49, 345-350.
37. Voutsinas, L. P.; Cheung, E; Nakai, S. J. Food Sci. 1983, 48, 26-32.
38. Marshall, K. R.; Harper, W. Bull. Int. Dairy Fed. 1988, 233, 21-32.
39. Lee, C-H.; Kim, S-K. Food Hydrocolloids 1987, 1(4), 283-289.
40. Kanno, C. J. Food Sci. 1989, 54(6), 1534-1539.
41. Shimizu, M.; Saito, M.; Yamauchi, K. Agric. Biol. Chem. 1985, 49(1), 189-194.
42. Shimizu, M.; Kamiya, T.; Yamauchi, K. Agric. Biol. Chem. 1981, 45(11), 2491-2496.
43. Yamauchi, K.; Shimizu, M; Kamiya, T. J. Food Sci. 1980, 45, 1237-1242.
44. Das, K. P.; Kinsella, J. J. Disp. Sci. Tech. 1989, 10(1), 77-102.
45. Kinsella, J. E.; Whitehead, D. In Advances in Food Emulsions and Foams; Dickinson, E; Stainsby, G., Eds.; Elsevier Applied Science: New York, 1988; pp 163-188.

46. Dickinson, E.; Stainsby, G., Eds. In Advances in Food Emulsions and Foams; Elsevier Applied Science: New York, 1988; pp 1-44.
47. Halling, P. J. CRC Crit. Rev. Food Sci. Nutr. 1981, 15(1), 155-203.
48. Waniska, R. D.; Kinsella, J. Food Hydrocolloids 1988, 2(1), 59-67.
49. Toro-Vazquez, J. F.; Regenstein, J. J. Food Sci. 1989, 54(5), 1177-1185, 2201.
50. Morr, C. V. In Protein Functionality in Foods; Cherry, J., Ed.; American Chemical Society: Washington, DC, 1981; pp 201-215.
51. Matsudomi, N.; Mori, H.; Kato, A.; Kobayashi, K. Agric. Biol. Chem. 1985, 49, 915-919.
52. Dickinson, E.; Rolfe, S.; Dalgleish, D. Food Hydrocolloids 1989, 3(3), 193-203.
53. Dickinson, E. J. Dairy Res. 1989, 56, 471-477.
54. Leman, J.; Haque, Z.; Kinsella, J. Milchwissenshaft 1988, 43(5), 286-288.
55. Fligner, K. L.; Mangino, M. Food Hydrocolloids 1990, submitted.
56. Fligner, K. L.; Mangino, M. Food Hydrocolloids 1990, submitted.
57. Nakamura, R.; Mizutani, R.; Yano, M.; Hayakawa, S. J. Agric. Food Chem. 1988, 36, 729-732.
58. Dickinson, E.; Stainsby, G. Food Tech. 1987, 9, 74-81, 116.
59. Tolstoguzov, B. V. In Functional Properties of Food Macromolecules; Mitchell, J.; Ledward, D., Eds.; Elsevier Applied Science Publishers: New York, 1986; pp 385-415.
60. Tolstoguzov, V. B. Food Hydrocolloids 1988, 2(5), 339-370.
61. Jimenez-Flores, R.; Richardson, T. J. Dairy Sci. 1988, 71(10), 2640-2654.
62. Arai, S.; Watanabe, M. In Advances in Food Emulsions and Foams; Dickinson, E.; Stainsby, G., Eds.; Elsevier Applied Science: New York, 1988; pp 189-220.
63. Mitchell, J. R. In Developments in Food Proteins-4; Hudson, B., Ed.; Elsevier Applied Science Publishers: New York, 1986; pp 291-338.
64. Poole, S.; Fry, J. In Developments in Food Proteins-5; Hudson, B., Ed.; Elsevier Applied Science: New York, 1987; pp 257-298.
65. Bacon, J. R.; Hemmant, J.; Lambert, N.; Moore, R; Wright, D. Food Hydrocolloids 1988, 2(3), 225-245.
66. Clark, D. C.; Mackie, A.; Smith, L.; Wilson, D. Food Hydrocolloids 1988, 2(3), 209-223.
67. Phillips, L. G.; Davis, M.; Kinsella, J. Food Hydrocolloids 1989, 3(3), 163-174.
68. Volpe, T.; Zabik, M.; Cereal Chem. 1975, 52, 188-197.
69. Anderson, M.; Brooker, B. In Advances in Food Emulsions and Foams; E. Dickinson, E.; Stainsby, G., Eds.; Elsevier Applied Science: New York, 1988; pp 221-255.
70. Lorient, D.; Closs, B; Courthaudon, J. J. Dairy Res. 1989, 56, 495-502.
71. Poole, S. Int. J. Food Sci. and Tech. 1989, 24, 121-137.
72. Phillips, L. G.; Yang, S.; Schulman, W.; Kinsella J. J. Food Sci. 1989, 54(3), 743-747.
73. Poole, S.; West, S.; Fry, J. Food Hydrocolloids 1987, 1(3), 227-241.
74. Hansen, P. M. T.; Black, D. J. Food Sci. 1972, 37, 452-456.

RECEIVED June 20, 1990

Chapter 2

Significance of Macromolecular Interaction and Stability in Functional Properties of Food Proteins

Akio Kato

Department of Agricultural Chemistry, Yamaguchi University, Yamaguchi 753, Japan

Beside protein hydrophobicity, the role of macromolecular interaction and conformational stability in the functional properties of proteins are described here. The importance of protein association in the foaming properties was confirmed using various polymeric forms of ovomucin or heat-induced ovalbumin aggregates, while electrostatic repulsion enhancing protein dissociation was indicated to be an important factor in the emulsifying properties of proteins using the polyanionic proteins, ovomucin and phosvitin. The role of conformational stability of proteins in the surface properties was proved by a protein engineering approach using several mutant tryptophan synthases with different stabilities which were obtained by a single amino acid substitution at position 49. Good correlations were observed between the conformational stability and the surface properties of mutant tryptophan synthases.

Recently many attempts to develop new functional food proteins have been done by food researchers to meet the world's protein needs. At the same time, since it is of importance to elucidate the molecular basis of protein functionality for developing new functional proteins, many food chemists have focused their efforts on elucidating the relations between the structural and functional properties of proteins. Of the functional properties of food proteins, much emphasis has been placed on surface properties such as foaming and emulsifying properties because they are critical elements in the development of new food proteins. If we can know the relationships between the structural and functional properties, it would be possible to make an educated design of new functional proteins by means of various protein modification including protein engineering. However, our understanding of the determinants of protein functional properties is still limited. We have proposed (1) that the surface hydrophobicity of proteins is the important structural factor

0097–6156/91/0454–0013$06.00/0

governing functional properties. Many studies (2-5) have supported
our hypothesis. In addition to surface hydrophobicity of proteins,
the relation of protein association/dissociation and protein
stability to the functional properties of food proteins should be
studied. However, these structural and functional properties are
difficult to define because the structural properties of proteins
compared in many experiments, namely, molecular size, net charge,
stability and surface hydrophobicity, were markedly different from
one another. In this paper we use model proteins to get clear-cut
structural and functional relationship. First, effects of protein
association and dissociation on the emulsifying and foaming
properties were investigated using various polymeric forms of
ovomucin and heat-induced soluble aggregates of ovalbumin. Second,
roles of electrostatic repulsion enhancing protein dissociation in
the functional properties of proteins were studied using polyanionic
proteins, ovomucin and phosvitin. Finally, significance of protein
stability on the functional properties was investigated using mutant
tryptophan synthases prepared by protein engineering procedure. It is
strongly emphasized here that protein engineering is a very powerful
approach to interpretation of the structural and functional
properties of food proteins.

Materials and Methods

Ovomucin was prepared by the method described in a previous paper(6).
Ovomucin gel was solubilized in 1/20 M Tris-HCl buffer at pH 8. The
soluble ovomucin was treated with a sonicator at 5 A for 5 min at 20
°C. The sonicated ovomucin was further reduced with 0.01 M
mercaptoethanol.

Ovalbumin was crystallized with sodium sulfate from fresh egg
white and recrystallized five times. The heat-induced soluble
aggregates of ovalbumin were formed by heating to 80 °C in 67 mM
phosphate buffer at pH7 containing 0.1 M NaCl with an incubator at a
rate of 3°C/min and then immediately cooled at room temperature for
0, 3 and 6 hrs to get different molecular weights of aggregates. The
molecular weights of soluble aggregates were determined by the low-
angle laser light scattering technique combined with high performance
gel chromatography (7).

Phosvitin was prepared by the method of Mecham and Olcott (8)
and further purified by ion exchange chromatography on a DEAE-
Sephadex A-50.

The wild-type tryptophan synthase α-subunits were obtained by
the method of Yutani et al.(9). Six mutant proteins substituted by
alanine, glycine, isoleucine, lysine, phenylalanine and threonine in
place of glutamic acid at position 49 were obtained by site-directed
mutagenesis using synthetic oligonucleotides(10). The single-stranded
DNA template used for the mutagenesis was an M13mp11 derivative
containing the α-subunit gene isolated from an HincII fragment(5632-
6925) of tryptophan operon. A plasmid derived from plasmid pUC8
(Pharmacia) containing the tryptophan promoter introduced from pDR720
(Pharmacia) was used as the expression vector. Purified mutants
(20-100 mg) were obtained from 5 liters of broth.

A quantitative method is required to study the relationship
between the structural and functional properties of proteins. Pearce
and Kinsella(11) have developed the turbidity method to measure

the emulsifying properties. We have developed a conductivity method
to estimate the foaming properties (12). These methods were used to
determine the emulsifying and foaming properties, because of their
reliable and quantitative yields.

Results and Discussion

Effects of Protein-Protein Interaction on Foaming Properties. The
role of protein-protein interaction in surface functional properties
is first studied using various polymeric forms of ovomucin. Ovomucin
is a glycoprotein responsible for gel structure of thick egg white.
And it is well known as the best foaming protein among natural
proteins. Soluble ovomucin is a polymeric form (8300 KDa) of
heterogeneous subunits consisting of α- and β-ovomucin, and is
dissociated with sonication or reduction into smaller units of
molecular weight of 1100 KDa and 540 KDa, respectively, without
changes in chemical composition (13). As shown in Figure 1, good
foaming properties are observed in the highly polymerized form of
ovomucin, while it decreases in the dissociated forms, depending on
the molecular weight. On the other hand, the emulsifying properties
of ovomucins showed a good correlation with the surface hydrophobicity, not with the polymerization(14).

     Similarly, various polymeric forms of ovalbumin aggregates were
used to study the effect of protein association on the surface
properties. When ovalbumin is heated at controlled conditions (at 80
°C, pH 7.4), it forms soluble aggregates. The average molecular
weights are increased with prolonged standing at room temperature
after heat-treatment(7). Figure 2 shows the foaming properties of
soluble ovalbumin aggregates heated at 80°C and then cooled at room
temperature for 0, 3, 6 hr whose molecular weights are 6,000,000,
7,800,000 and 9,500,000, respectively. The foaming properties
increased with increasing molecular weight of the soluble aggregates.
In contrast, the emulsifying properties were found to be unaffected
(data are not shown here). From these results, it is suggested that
protein association is an important structural factor governing the
foaming property, while the emulsifying properties are hardly related
to such events.

Role of electrostatic repulsion in the emulsifying properties.Illus-
tration of the role of protein-protein interaction in the surface
properties was also achieved when polyanion-type proteins were used
which inhibit the interaction at the interface. Ovomucin and
phosvitin were used as polyanion-type proteins. Ovomucin has about
10 % sialic acids in the terminal carbohydrate chains, and phosvitin
have about 30 % phosphate bound to serine residues. These terminal
polyanions can easily be removed by enzymatic action without affecting peptide bonds. Figure 3 shows the effect of terminal sialic acid
residues on the emulsifying property of ovomucin (15). The terminal
sialic acid was removed by sialidase. Removal of sialic acids greatly
affected the functional properties. The emulsifying properties of
ovomucin were greatly decreased by the removal of sialic acid,
suggesting that sialic acid plays an important role for the
emulsifying properties. On the other hand, the foaming properties of
ovomucin were increased by the removal of sialic acid, as shown in a
previous paper(15). This may be because the decrease in negatively

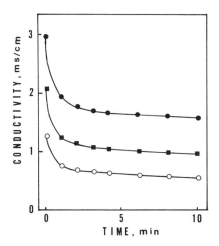

Figure 1. Foaming properties of native(●), sonicated(■) and reduced(O) ovomucins

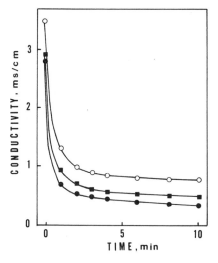

Figure 2. Foaming properties of the soluble aggregates of ovalbumin heated to 80 C and then cooled at room temperature for 0 (●), 3 (■) and 6 (O) hours

charged sialic acids enhances the interaction of ovomucin molecules at air-water interface, thus leading to strengthening of the foam film. Figure 4 shows the effect of phosphate residues on the emulsifying properties of phosvitin (16).The emulsifying properties were greatly decreased by removal of phosphate by phosphatase. The foaming properties of phosvitin were too poor to detect any changes. Thus, removal of sialic acid and phosphate from ovomucin and phosvitin, respectively, greatly affected their functional properties. The foaming properties of these proteins were promoted by the removal of polyanions, while the emulsifying properties were decreased. Accordingly, these results illustrate the importance of electrostatic repulsive forces in the stabilization of emulsion droplets. These observations run parallel with our proposal that surface hydrophobicity is important in the emulsifying properties of proteins. The surface hydrophobicity may act as the major driving force for the emulsion formation and the electrostatic repulsion may serve to stabilize the formed emulsion by preventing the coalescence of emulsion oil-droplets.

Additional evidence regarding the importance of electrostatic repulsion in the emulsifying properties of proteins has come from the fact that the emulsifying properties of proteins were improved by deamidation (17-19). Deamidation of proteins results in an increase in negative charges which result from the hydrolysis of amide groups in glutamine and asparagine. The effects of deamidation on the emulsifying properties are especially remarkable for wheat gluten containing a high amount of glutamine and asparagine(19). Thus, the effects of electrostatic repulsion on the emulsifying properties of proteins seems to be fairly drastic.

Significance of protein stability in the functional properties.
Protein stability seems to be an important factor influencing surface properties because it is generally accepted that unstable protein molecules unfold with hydrophilic segments oriented toward the aqueous phase and hydrophobic segments oriented toward the air or oil phase to cause the pronounced reduction of surface or interfacial tension that facilitates foaming and emulsification. To prove this hypothesis, we have employed a protein engineering approach. The effects of single amino acid substitutions on the stability of proteins have been studied using site-directed mutagenesis. In this respect, a series of variant α-subunits of tryptophan synthase substituted by each of 20 amino acids at position 49 have been constructed by Yutani et al.(10). In the case of the tryptophan synthase α-subunit substituted at position 49, the Gibbs free energy of unfolding in water of various mutants, which is a quantitative measure of protein stability, has varied between 0.72 and 1.92 times that of wild-type proteins. Therefore, they formed an ideal set of mutants with which it would be fruitful to estimate the relation between the surface properties and conformational stability. Fortunately, a large amount of mutants (20-100 mg per 5 liters of broth) are obtained (10). For this reason, we have studied (20) the relationship between the conformational stability and functional properties of proteins using typical mutant tryptophan synthase of varying stabilities. Table 1 shows the Gibbs energy of unfolding of wild-type and six mutant proteins. The free energy of unfolding is a function of the equilibrium constant (K) that relates the

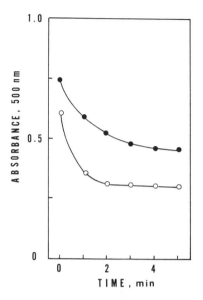

Figure 3. Effects of sialidase treatment on the emulsifying properties of ovomucin. ●,untreated; ○,treated with sialidase (Reprinted with permission from ref. 15. Copyright 1987 Japan Society for Bioscience, Biotechnology, and Agrochemistry.)

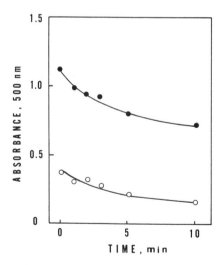

Figure 4. Effects of phosphatase treatment on the emulsifying properties of phosvitin. ●, untreated; ○, treated with phosphatase (Reprinted with permission from ref. 16. Copyright 1987 Japan Society for Bioscience, Biotechnology, and Agrochemistry.)

concentration of folded and unfolded forms ($\Delta G = - RTlnK$), and
reflects the stability of proteins. The stability of the mutant
proteins substituted by isoleucine in place of glutamic acid at
position 49 is about twice that of the wild-type at pH 7 and 9.
Figure 5 shows the foaming properties of wild-type and mutant α-
subunits of tryptophan synthase substitution at the same position 49
by site-directed mutagenesis. Since the α-subunit of tryptophan
synthase consists of 268 amino acid residues without a disulfide bond,
it is a very flexible protein and shows good foaming and emulsifying
properties. Despite only a single amino acid substitution at the same
position, a big change in the foaming properties associated with a
concomitant shift in the conformational stability was observed
between wild-type and mutant proteins. Thus, the relationships
between the structural and functional properties of the α-subunit of
tryptophan synthase were investigated using wild-type and 6 mutant
proteins substituted at position 49. As shown in Figure 6, good
correlations were observed between the surface tension and the ΔG of
unfolding. This suggests that the surface tension of proteins closely
depends on the conformational stability. Figures 7 and 8 show the
relationships between the ΔG of unfolding, the foaming power, and the
foam stability of mutant proteins, respectively. As clearly shown in
figures, good correlations were obtained between the ΔG for both
foaming power and foam stability. These data indicate that the
foaming properties of proteins are related to the conformational
stability. Figure 9 shows the relationship between the ΔG and
emulsifying activity of mutant proteins. A good correlation is
observed between the ΔG and emulsifying activity. The correlation
coefficients of the surface functional properties with the stability
(ΔG) of mutants of tryptophan synthase at pH 9 are summarized in
Table 2. The correlation coefficients of ΔG with foaming properties
were higher than those with emulsifying properties. Thus, the
conformational stability (ΔG) is confirmed to be the important
determinant for the functional properties, especially for foaming.
     Although the relationships between the surface hydrophobicity and
the surface properties of mutant tryptophan synthases were also
studied, their correlations were poor. A number of researchers (1-5)
have reported good correlations between the surface hydrophobicity
and the surface properties of proteins. It is reasonable to assume
that the surface hydrophobicity of proteins plays an important role
in emulsification and foam formation, because amphiphilic proteins
possessing high surface hydrophobicity are strongly adsorbed at the
interface between oil or air and water causing a pronounced reduction
of interfacial or surface tension that readily facilitates
emulsification and foaming. The results presented here delineate
unequivocally that the conformational stability of proteins should be
also considered as an important factor governing the surface
properties of unstable globular proteins. Since the values for the
surface hydrophobicity of proteins in solution may be different from
those at the oil/water interface, it is not enough to elucidate the
relationships between the structural and emulsifying properties using
only surface hydrophobicity as a structural factor. Therefore, it
should be more rational to consider the conformational stability as
another structural factor governing the emulsifying properties of
proteins.

TABLE 1. Values of Unfolding Gibbs Energy (ΔG) for Seven
α-Subunits of Tryptophan Synthase Substituted at Position
49 by Site-Directed Mutagenesis

Residue at position 49	ΔG	
	pH 7.0	pH 9.0
Glu(wild)	8.8 ± 0.1	4.9 ± 0.3
Gly	7.1 ± 0.1	6.4 ± 0.1
Ala	8.5 ± 0.2	6.8 ± 0.2
Ile	16.8 ± 0.5	10.0 ± 0.3
Phe	11.2 ± 0.2	8.3 ± 0.2
Lys	7.9 ± 0.5	7.5 ± 0.9
Thr	8.8 ± 0.2	7.0 ± 0.8

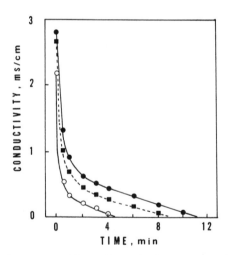

Figure 5. Foaming properties of typical mutants of tryptophan
synthase substituted at position 49.   O , substituted with
isoleucine; ■ , substituted with glycine; ● , wild-type (glutamic
acid)

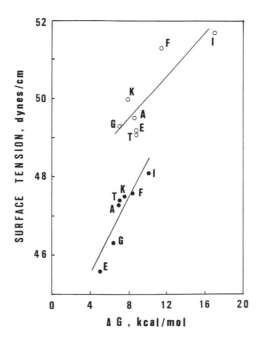

Figure 6. Relationship between ΔG and surface tension of
mutant tryptophan synthases substituted at position 49 at
pH 7 (O) and pH 9 (●). A, E, F, G, I, K and T indicate the data
of proteins whose amino acids at 49 are alanine, glutamic acid,
phenylalanine, glycine, isoleucine, lysine and threonine,
respectively.
(Reprinted with permission from ref. 20. Copyright 1988 Oxford University Press.)

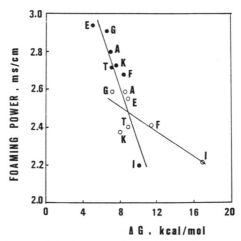

Figure 7. Relationship between ΔG and foaming power of mutant
tryptophan synthases substituted at position 49 at pH 7 (O)
and pH 9 (●). A-T, as in Figure 6.
(Reprinted with permission from ref. 20. Copyright 1988 Oxford University Press.)

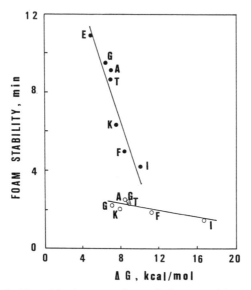

Figure 8. Relationship between ΔG and foam stability of mutant tryptophan synthases substituted at 49 at pH 7 (O) and pH9 (●). A-T, as in Figure 6.
(Reprinted with permission from ref. 20. Copyright 1988 Oxford University Press.)

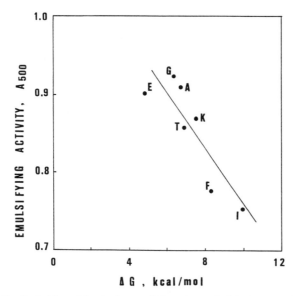

Figure 9. Relationship between ΔG and emulsifying activity of mutant tryptophan synthases substituted at 49 at pH 9.   A-T, as in Figure 6.
(Reprinted with permission from ref. 20. Copyright 1988 Oxford University Press.)

Relationship between structural and functional properties of proteins. Table3 summarizes the relationships between structural and functional properties of proteins which have been elucidated in this paper. The emulsifying properties are distinguished between emulsifying activity and emulsion stability. Both the surface hydrophobicity and the conformational stability of proteins are involved in the emulsifying activity. Beside these structural factors, the electrostatic repulsive force acting as an inhibitor of the coalescence of oil-droplets should be added to get stable emulsions. Similarly, foaming properties are distinguished between foaming power and foam stability. The main structural factor of foaming power is the conformational stability of proteins. There is no correlation between the foaming power and the surface hydrophobicity of proteins used here. To get a stable foam, proteins have to interact with themselves to strengthen foam film. Thus, protein-protein interaction is an important determinant for foam stability.

Conclusions The structural determinants of emulsifying and foaming properties of proteins have been elucidated. As the relationships between structural and functional properties become clearly established, we will be able to design more readily functional proteins using protein engineering techniques. Current and future advances in the genetic engineering of food proteins should make possible the production of new functional proteins, improving their nutritional or other functional properties. As shown in this paper, the genetic alteration of amino acid residues can dramatically alter the functional properties of proteins, even at the level of a single amino acid substitution. This suggests the possibility of the development of new functional proteins by protein engineering approach.

TABLE 2.   Correlation Coefficients of Surface Properties at
            pH 9 with the ΔG of Tryptophan Synthase α-Subunits

Surface properties	r
Surface tension	0.90 ($p < 0.01$)
Foaming power	−0.93 ($p < 0.01$)
Foam stability	−0.95 ($p < 0.01$)
Emulsifying activity	−0.83 ($p < 0.05$)

TABLE 3.   Relationships between Structural and Functional
            Properties of Proteins

Functional properties	Structural factors
Emulsifying properties	
emulsifying activity	surface hydrophobicity, protein stability
emulsion stability	electrostatic repulsive force in hydrophilic site
Foaming properties	
foaming power	protein stability
foam stability	protein-protein interaction

Literature Cited

1. Kato,A.; Nakai,S. Biochim. Biophys. Acta 1980,624, 13-20.
2. Nakai,S. J. Agric. Food Chem. 1983, 31, 676-683.
3. Shimizu,M.; Takahashi,T.; Kaminogawa,S.; Yamauchi,K. J. Agric.
   Food Chem. 1983, 31, 1214-1218.
4. Voutsinas,L.P., Cheung,E.; Nakai,S. J. Food Sci. 1983, 48, 26-32.
5. Townsend,A.A.; Nakai,S. J. Food Sci. 1983, 48, 588-594.
6. Kato,A.;Nakamura,R.; Sato,T. Agric. Biol. Chem. 1970, 34, 854-859.
7. Kato,A.; Takagi,T. J. Agric. Food Chem. 1987, 35, 633-637.
8. Mecham,D.K.; Olcott,H.S. J. Am. Chem. Soc. 1949, 71, 3670-3678.
9. Yutani,K.;Ogasahara,K.;Aoki,K.;Kakuno,T.;Sugino,Y. J. Biol. Chem.
   1984, 259, 14076-14081.
10. Yutani,K.;Ogasahara,K.;Tsujita,T.;Sugino,Y. Proc. Natl. Acad. Sci.
    USA 1987, 84, 4441-4444.
11. Pearce,K.N.; Kinsella,J.E. J. Agric. Food Chem. 1978, 26,
    716-723.
12. Kato,A.;Takahashi,A.; Matsudomi,N.; Kobayashi,K. J. Food Sci.
    1983, 48 , 62-65.
13. Hayakawa,S.; Sato,Y. Agric. Biol. Chem. 1976, 40, 2397-2404.
14. Kato,A.; Oda,S.; Yamanaka,Y.; Matsudomi,N. Kobayashi,K. Agric.
    Biol. Chem. 1985, 49, 3501-3504.
15. Kato,A.; Miyachi,N.;Matsudomi,N.; Kobayashi,K. Agric. Biol. Chem.
    1987, 51, 641-645.
16. Kato,A.;Miyazaki,S.;Kawamoto,A.;Kobayashi,K. Agric. Biol. Chem.
    1987, 51, 2989-2994.
17. Wu,C.H.;Nakai,S.;Powrie,W.D. J. Agric. Food Chem. 1976, 24,
    504-510.
18. Kato,A.; Tanaka,A.; Lee,Y.; Matsudomi,N.; Kobayashi,K. J. Agric.
    Food Chem. 1987, 35, 285-288.
19. Matsudomi,N.; Tanaka,A.; Kato,A.; Kobayashi,K. Agric. Biol. Chem.
    1986, 50, 1989-1994.
20. Kato,A.; Yutani,K. Protein Engineering 1988, 2, 153-156.

RECEIVED June 20, 1990

Chapter 3

Effect of Preheat Temperature on the Hydrophobic Properties of Milk Proteins

N. Parris, J. H. Woychik, and P. Cooke

Eastern Regional Research Center, Agricultural Research Service, U.S. Department of Agriculture, Philadelphia, PA 19118

Pre-heat treatment of skim milk for the preparation of nonfat dry milk (NDM) powders results in physicochemical changes that effect the functional behavior of the powder. The functional properties of NDM powders heated to 63°C (I), 74°C (II), and 85°C (III) for 30 min before spray drying were different. Comparison of reversed phase chromatographic profiles indicated that milk proteins were altered and a whey-casein complex was formed in II and III. Hydrophobic interaction chromatography (HIC) indicated that the complex formed eluted before ß-casein and hence was more hydrophilic than was indicated under reversed phase chromatography conditions. Alkane binding to the milk proteins in the rehydrated powders also indicated that II and III were more hydrophilic than I. Electron micrographs of immuno-gold labeled whey proteins in the rehydrated powders indicated that less than ten percent of the gold particles were associated with the casein micelles for I but greater than fifty percent were associated with the micelles for II and III. The appearance of the complex formed in II and III on the surface of casein micelles suggests it may contribute to the increased hydrophilic character of these powders. Examination of their functional properties indicated that III had the lowest solubility and the greatest percent overrun in a foaming test. The emulsifying activity index (EAI) was greatest for II and no significant differences were observed for foam and emulsion stability between the three powders.

Nonfat dry milk (NDM) is frequently used as a functional
ingredient for dairy, bakery, confectionery, and other
food applications. Heat treatment of skim milk before
spray drying is used widely as a means of manipulating
the functional properties of NDM powders in milk products. As a result, the end-use of such products is very
dependent on the heat treatment the powders receive (1).
Heat treatments can denature whey proteins resulting in
aggregation which alter the ability of the powders to
rehydrate. Although difficult to distinguish, aggregation is a separate and usually irreversible process which
follows denaturation of whey protein. Denaturation is
normally reversible and can be stopped before aggregation
begins (2). Limited whey protein denaturation can improve
the emulsifying properties of whey protein in food systems (3). Caseins, which account for about 80% of the
total milk proteins, exist in milk as soluble complexes
or micelles. Caseins are amphiphilic in nature and are
primarily responsible for the excellent surfactant properties or functionality of milk ingredients. Unlike whey
proteins the caseinate system in milk is very stable to
heat and tends to resist coagulation. Heat treatment,
however, is known to increase the acidity of milks,
primarily through the decomposition of lactose, and also
to reduce both total soluble and ionic calcium resulting
in flocculation of casein micelles and a corresponding
loss of functional properties.
    Preheat treatment of skim milk between 85°C and
100°C for 30 min in the preparation of NDM powders is
commonly used in the baking industry to improve the
extensibility and water absorption of dough (4). High-
heat NDM powders (85°C) absorb more water than lower
heated powders and this can improve its emulsion stability and gelation properties (5). Such functional properties are important in comminuted meat, confectionery,
as well as in reconstituted concentrated sterile and
baked products. Generally extensive denaturation is used
to prevent undesired side reactions such as loaf depres-
sion or coagulation. Because some water absorption is
necessary, ice cream and dairy beverages are frequently
made with low- and medium-heat NDM (63°C, 74°C). Yogurt
texture is greatly affected by the degree of whey protein
denaturation (6). Batch-type heating of skim milk between 85-95°C for 5 to 10 min before inoculation is an
important processing variable in the manufacture of
yogurt.
    The purpose of this study was to assess the heat
induced physicochemical changes that occur during manufacturing of NDM powders. Changes in milk protein
profiles were identified using chromatographic methods

and micellar associations of whey proteins or complexes
were examined by immuno-labeled electron microscopic
techniques.

Material and Methods

Nonfat Dry Milk. Raw pooled herd milk from Holstein,
Ayrshire, and Brown Swiss cows was skimmed at 39°C and
preheated at 63°C, 74°C, and 85°C for 30 min; concentrated; then spray dried to yield NDM powders I, II, and
III according to a previously published procedure (7).

Hydrophobic Interaction Chromatography (HIC).
Chromatographic separations of milk proteins were carried
out on a Spectra Physics (San Jose, CA) SP-8800 HPLC
system, equipped with a LC-HINT column, 0.46 x 10 cm,
(Supelco, Inc., Bellefonte, PA); mobile phase, solvent B:
0.05M sodium phosphate, pH 6.0, in 3.75M urea, solvent A:
2.0M ammonium sulfate in solvent B; gradient, 0%-100%B,
30 min; flow rate, 0.8 mL/min; detector gain 0.1 aufs.
Sample, 1.0 g NDM was dissolved in 10 mL of water, and
centrifuged at 100,000 x g, 4°C, 40 min. The supernatant
was filtered (0.45 μM pore size) and 200 μL injected onto
the column. The pellet, 3 mg, was dissolved in 1 ml of
solvent B, H_2O, solvent A, separately, and 100 μL injected onto the column.

Gel Electrophoresis. SDS- and urea-PAGE of skim milk
proteins was carried out on a PhastSystem (Pharmacia,
Piscataway, NJ) as previously described (8) and Van
Hekken, D. et al. J. Dairy Sci. in press). The molecular
weight standards from Bio-Rad, (Richmond, CA) and their
corresponding molecular weights for SDS-PAGE were: phos-
phorylase b, 97,400; bovine serum albumin (BSA), 66,200;
ovalbumin, 42,699; carbonic anhydrase, 31,000; soybean
trypsin inhibitor, 21,500; lysozyme, 14,000.

Electrophoretic Blotting. A modification of the method
of Towbin et al. (9) was used for the electrophoretic
transfer of milk proteins from the polyacrylamide gel to
nitrocellulose (NC) membrane. Half of the gel containing
the skim milk proteins (39°C) was stained with Coomassie
R350 dye and the other half was removed from the cello-
phane backing for electroblotting. The gel isolated for
immunoblotting was allowed to equilibrate in transfer
buffer; 25mM Tris, 192mM glycine, and 20% methanol, pH
8.3, for 5 min. The proteins were transferred to a NC
membrane using a Mini Trans-Blot cell (Bio-Rad, Richmond,
CA) at 60 V, 0.15 A, for 3 h at room temperature with
electrodes 8 cm apart.

Immunological Detection. To avoid nonspecific binding,
the NC membrane was incubated in blocking buffer; 50mM
Tris, 150mM NaCl, pH 7.4, containing 2.5 g gelatin and

2.5 mL 10% NP-40. The membrane was incubated for 1 h
with primary immune antiserum to cow whey protein Cat.
#AXL-306 (Accurate Chemical and Scientific Corp.,
Westbury, NY) and secondary antibody, goat antirabbit
IgG conjugated to horse radish peroxidase (Behring
Diagnostic, La Jolla, CA) for 1 h. Both antisera were
diluted 1:1000 in blocking buffer. The peroxidase substrate was developed in 4-chloro-1-naphthol, 10 mg in
3.3 ml methanol 16.7 ml blocking buffer and 10 μL of 30%
H_2O_2.

Electron Microscopy. Samples of heated skim milk and
rehydrated NDM powders were prepared separately for
immunolabeling as whole mounts by negative staining and
as thin sections, embedded in plastic. For labeling with
negative staining, aliquots of milk containing 0.1%
gelatin were incubated for 1 h at room temperature with
rabbit antiserum to cow whey protein in phosphate-
buffered saline (PBS). The mixture was absorbed to
Formvar-carbon coated Ni Specimen grids for 10 min,
washed with 5-10 drops of 1% gelatin in PBS, and incubated for 1 h on a 100 μL drop of goat antirabbit IgG,
conjugated to 10 nm colloidal gold particles GAR 10,
(Jannsen Life Sciences Products, Beerse, Belgium). The
samples were washed sequentially with: 1% gelatin in PBS,
PBS, water, and finally stained with 2% uranyl acetate
solution.
     For immunolabeling of thin sections, fresh samples
of milk were embedded in the low temperature embedment,
Lowicryl K4M, according to the procedure of Altmann et
al. (10). Thin sections on Formvar-carbon coated Ni
specimen grids were incubated sequentially with: 1%
gelatin in PBS, rabbit antiserum to cow whey protein,
gelatin/PBS, and goat antirabbit IgG, conjugated to
colloidal gold particles. Sections were then washed with
gelatin/PBS, PBS, and water before post staining the
sections with 2% uranyl acetate.

Hydrophobicity. Protein hydrophobicity was determined by
measuring the extent of alkane binding to NDM protein by
a modification of the procedure of Mangino et al. (11).
Two mL of aqueous NDM solution (10%) and heptane (1.5 mL)
were added to 5 mL vials, sealed with parafilm, and mixed
sideways at 3 rev/min at 25.0 ± 0.2°C for 18 h with a
model R-7636-00 Roto-torque rotator, (Cole-Palmer, International Chicago, IL, USA). The alkane layer was discarded and the aqueous phase (1.0 ml) was extracted with
undecane (0.6 mL) containing an internal standard
(octane) for 2 h at 3 rev/min. Samples (0.5 μL) of the
undecane phase were injected into a model 5710A gas
chromatograph (Hewlett-Packard Co., Palo Alto, CA)
equipped with a flame-ionization detector. Separations
were carried out on a column of Chromosorb W coated with
10% (w/w) OV-101.

<u>Functional Properties.</u> Solubility measurements were
performed as previously described (7). Emulsions were
prepared by homogenizing 3 mL of protein solution (10%)
and 1 mL corn oil for 30 sec using a Polytron homogenizer
at setting #5 and determining the emulsion activity index
(EAI) by turbidity measurements at protein concentrations
of 0.1, 0.5 and 1.0% and pH values of 6, 7, 8, and 9
according to the procedure of Pearce and Kinsella (12).
Foam formation (% overrun) and foam stability were determined according to the procedure of Phillips et al. (13)
with the modification that foam stability was determined
by measuring the time required for drainage equal to 50%
of the weight of foam (approximately 100 mL).

Results and Discussion

<u>HPLC Separations.</u> Earlier results have shown that
elution profiles for soluble dialyzed material from
rehydrated medium- and high-heat NDM powders (II and III,
respectively) were significantly different from the
low-heat powder (I) profiles (8). The profiles for I and
III are depicted together with the SDS-PAGE patterns of
the proteins present in corresponding peaks from the
analytical column (inset) in Figure 1. The whey-casein
complex formed in III was comprised of many proteins
including BSA, κ-, and α_{s2}-caseins, ß-lactoglobulin
(ß-LG), and α-lactalbumin (α-LA) (Figure 1b, lane 6).
Although RP-HPLC profiles and SDS-PAGE gels of reduced
proteins indicate that the complex is stabilized through
disulfide linkage, some of the native whey proteins
reformed during storage suggesting that whey complexes
are also present which are stabilized through less
specific interactions (hydrophobic or calcium-dependent
linkages) with the casein micelles. Comparison of
retention times for RP-HPLC separation of milk proteins
from III indicated that the whey-casein complex is more
hydrophobic under the conditions of chromatography than
the other milk proteins including ß-casein (Figure 1b).
   Elution of milk proteins from the same rehydrated
powders using hydrophobic interaction chromatography
(HIC) indicated that the whey-casein complex was less
hydrophobic than ß-casein. The whey-casein complex was
not present in the supernatant from I but was present in
the elution profile of the supernatant from III and
eluted before ß-casein (Figure 2). The complex was also
found in the pellet from III and eluted before the
caseins (Figure 3b). SDS-PAGE patterns of the whey-
casein complex (left inset) Figure 3b, lane 1, indicated
that the sample contained aggregates that did not enter
the running gel. Casein micelles apparently were not
completely dissociated in the HIC buffer since ß- and
α_{s1}-caseins were present in the complex fraction along
with κ-casein and the whey proteins (see identification
of milk proteins Figure 1a). The second and third peaks

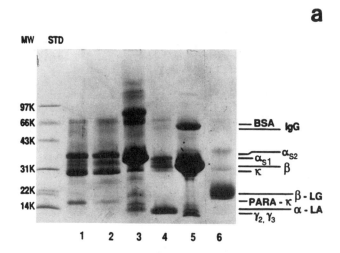

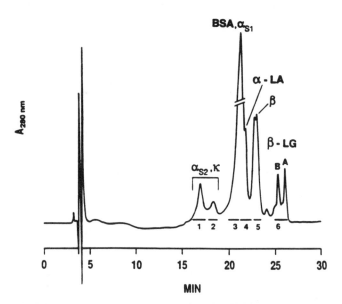

Figure 1. RP-HPLC elution profile and electrophoretic (SDS-PAGE) patterns of dialyzed soluble proteins from rehydrated NDM. a. I, b. III. (Reprinted from ref. 8. Copyright 1990 American Chemical Society.)

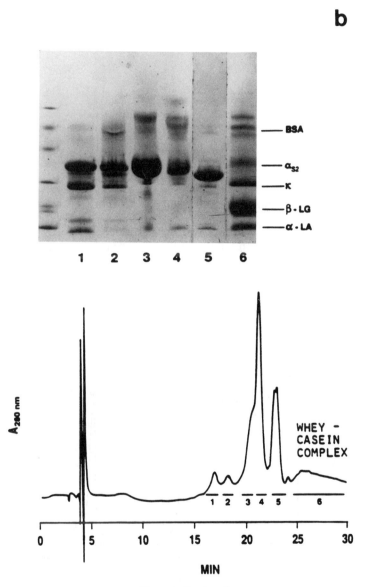

Figure 1. Continued.

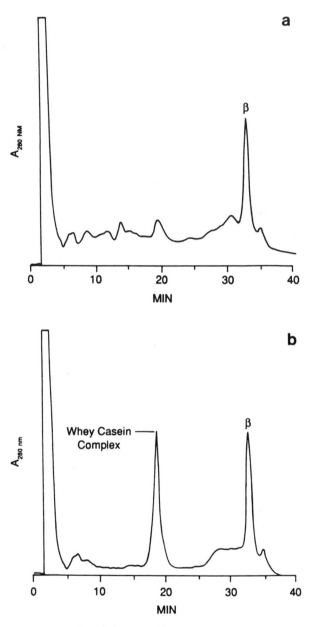

Figure 2. HIC elution profile of supernatant from rehydrated NDM powders. a. I, b. III.

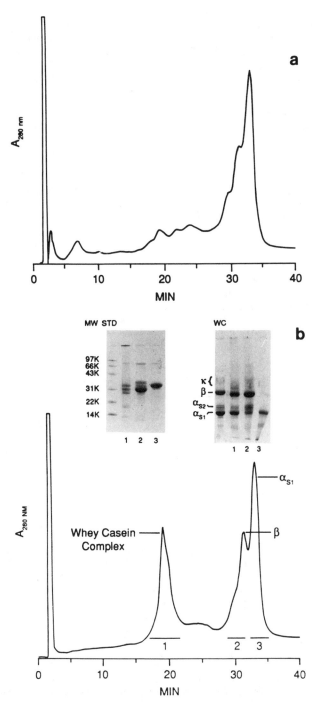

Figure 3. HIC elution profile and electrophoretic patterns of pellet from rehydrated NDM powders. a. I; b. III (inset, left = SDS-PAGE, right = urea-PAGE).

in the same figure were composed primarily of ß- and
α_{s1}-caseins respectively (left) inset lanes 2 and 3).
Caseins in whole casein (WC) can be identified more
easily with urea-PAGE (Figure 3, right inset). Urea-PAGE
of the whey-casein complex (right inset, lane 1) clearly
shows the presence of α_{s2}-casein, which was not detected
by SDS-PAGE, probably because of its poorer solubility in
that chromatographic system. Chromatographic conditions
used for HIC of the whey-casein complex formed on heating
skim milk before spray drying, therefore, indicate that
the complex is less hydrophobic than previously indicated
by RP-HPLC.

Heptane Binding. The amount of heptane bound to the
proteins for the three powders decreased with increasing
preheat temperature (Table I). Although the difference
between I and II was not statistically significant there
was a significant difference in alkane binding between
I and III. This could be attributed to the loss of
ß-lactoglobulin's alkane binding site to denaturation
however charged residues for most proteins are preferentially located at the protein surface, hydrophobic groups
buried within the protein away from its surface. Heat
denaturation should therefore expose more hydrophobic
residues; bind more alkane and improve functional behavior.

Table I. Alkane binding to rehydrated NDM powders

NDM	Heptane bound (mg/g protein)
LOW (I)	162 ± 9
MEDIUM (II)	144 ± 10
HIGH (III)	139 ± 4

n = 8

Location of Whey Protein Complexes. In order to better
understand the spacial relationship between the whey-
casein complex formed on heating milk and the casein
micelle, whey proteins in samples were immunolabeled with
colloidal gold and examined by electron microscopy. A
western blot of immune antiserum to cow whey proteins was
run against skim milk (39°C) to determine antibody specificity. SDS-PAGE separation of milk proteins is shown in
Figure 4 (left half) along with an electroblot of the
same proteins onto the NC membrane (right half). Results
indicated that immune antiserum bound to BSA and α-LA;
but binding to IgG, ß-LG and the caseins was weaker.
Electron micrographs of immuno-gold labeled, negative

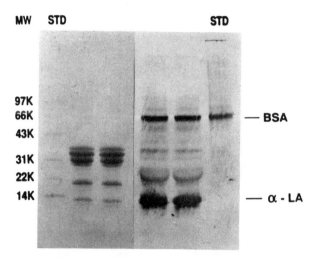

Figure 4. Western blot of antiserum to cow whey
protein against milk (39°C). Left half, SDS-PAGE Gel;
Right half, NC membrane

stained, heated skim milk samples indicated little or no
label associated directly with the surfaces of micelles
at 39°C, instead a few isolated gold particles were
located in spaces between micelles (Figure 5a). However,
(Figure 5b) the surfaces of many micelles from milks
heated at 85°C were labeled with several colloidal gold
particles. Label was primarily associated with material
on or near the surfaces of micelles (arrows), although
some isolated particles and small groups of gold label
were located in areas between micelles. For post-
embedment immuno-labeling of thin sections at 39°C,
(Figure 5c) some isolated particles of colloidal gold
label were located on or near the surfaces of micelles.
At 85°C, (Figure 5d) numerous particles of gold label
were found either singly or, more often, in clusters,
mostly in areas related to the (cut) surface of micelles
(arrows). Electron micrographs of immuno-gold labeled
negative stained rehydrated NDM powders, (Figure 6) indicated that most of the labeled particles were found in
the areas between micelles for I, whereas most of the
particles were associated with the irregular surface of
the micelles for II and III (arrows). To determine quantitatively the location of the gold particles in the
rehydrated milk samples, 100 particles were counted in
each of the negative stained preparations. Greater than
50% of the particles were associated with micelles for II
and III and less than 10% for I (see bar graph, Figure
6). Although several conflicting models have been proposed describing the structure of the casein micelle,
most current models put κ-casein on the casein micelle
surface. This model allows denatured serum proteins to
interact with κ-casein on the micelle surface with possible interconnection of the micelles. Our results
indicate that neither the heated skim milk (Figure 5d)
nor III (Figure 6c) show significant linking of micelles
through labeled whey proteins.

Functional Properties. The functional properties of the
rehydrated NDM powders that were examined were solubility, emulsifying activity index (EAI), overrun, foam and
emulsion stability. Isolated insoluble protein from
rehydrated NDM subjected to 3 different processing temperatures was relatively small, (between 2-6% of the total
protein in the powders) and was greatest for III. Insol-
ubles in III also contained more lactose which has been
suggested to interact with protein to form larger aggregates that coagulate (3). The EAI for the three powders,
measured at protein concentrations of 0.1, 0.5, and 1.0%
and at pH values of 6, 7, 8, 9, was greatest at 0.1% and
pH 9 (Figure 7). Regardless of concentration and pH
there was a consistent increase in emulsifying activity
for II followed by a significant decrease for III. Since
it was determined that approximately 50 and 95% of the
whey protein in II and III respectively was denatured,

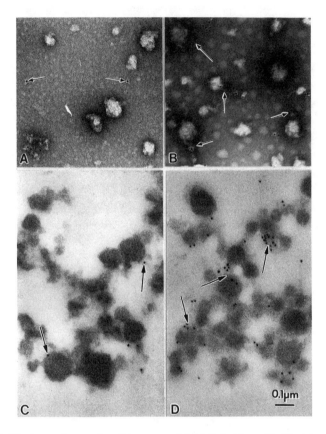

Figure 5. Electron micrographs of immuno-gold labeled skim milk. Heat treatment and preparation: A. 39 °C, negative stained; B. 85 °C, negative stained; C. 39 °C, thin sectioned; D. 85 °C, thin sectioned. Labeled whey protein (arrows).

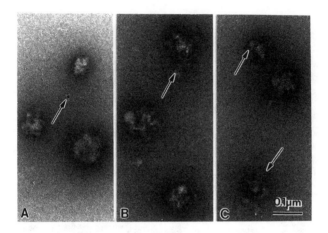

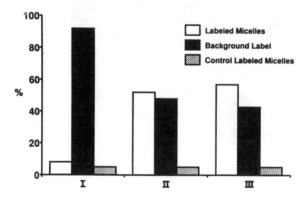

Figure 6. Electron micrographs of immuno-gold labeled rehydrated NDM powders. A. I, B. II, C. III; Negative stained. Labeled whey protein (arrows). Bar graph represents micellar association with particles (control contains no primary antibody).

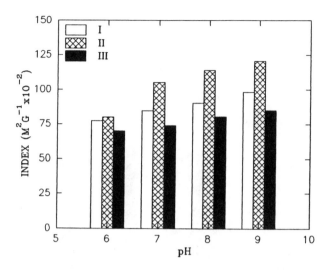

Figure 7. Emulsifying activity index for NDM powders 0.1% (w/v).

the greater EAI for II could be interpreted as the results of limited whey protein denaturation where unfolding of protein molecules exposes hydrophobic groups that can more easily orient at the oil-water interface and improve emulsifying properties. Mottar et al. ($\underline{6}$) have suggested that further heating results in a decrease in the ratio of ß-LG: α-LA associated with the casein micelle; an increase in water holding capacity and lower surface hydrophobicity. In general, mild heat treatment reduces whipping times, increases overrun, and enhances foam stability. For the rehydrated powders the overrun formed on whipping was consistently greater for III (Figure 8) however, little difference in foam stability was observed between the three powders (Figure 9).

Conclusions. This study demonstrates that preheat treatment of skim milk alters the micelle surface and effects the hydrophobic properties of NDM powders. Using various chromatographic and electron microscopic techniques the formation, composition, and location of the whey-casein complex in heated skim milk was elucidated. Of the functional properties evaluated, the emulsifying activity was the most interesting. Regardless of protein concentration and pH, the EAI was consistently higher for II. Although limited denaturation of milk proteins has been suggested to explain the greater emulsifying ability of II compared to I or III, further research is required to demonstrate that the increase is a result of protein localization at the oil-water interface. In addition, the embedding procedure chosen for EM was selected for

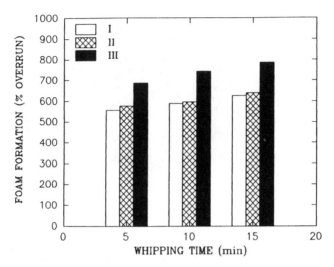

Figure 8. Foam formation (% overrun) for NDM powders.

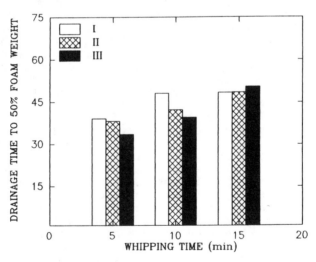

Figure 9. Foam Stability of NDM powders.

structural preservation and enhanced immunolabelling of the samples however the location of κ-, and α_{s2}-caseins in the micelle of heated and unheated milks was not determined. These two research areas warrant further investigation.

Acknowledgment

The authors thank Dr. M. Reinhart, Messrs. P. Smith, S. Ptashkin, T. Dobson and Mrs. D. Lu.

Literature Cited

1.   Pallansch, M. J.   In By-Products from Milk; Webb, B. H., Whittier, E. O., Eds.; AVI Publishing Co.: Westport, CT, 1970; Chapter 5.
2.   Brown, R. J.   In Fundamentals of Dairy Chemistry; Wong, N. P., Ed.; Van Nostrand Reinhold Co: New York, 1988; Chapter 11.
3.   Morr, C. V.   In Functionality and Protein Structure; Pour-El, A., Ed.; ACS Symposium Series No. 92; American Chemical Society: Washington, DC, 1979; pp 65-79.
4.   Guy, E. J.   In By-Products from Milk; Webb, B. H., Whittier, E. O., Eds.; AVI Publishing Co.: Westport, CT, 1970; Chapter 7.
5.   Kinsella, J. E.   CRC Crit. Rev. Food Sci. Nutr. 1984, 21, 197-262.
6.   Mottar, J.; Bassier, A.; Joniau, M.; Baert, J.   J. Dairy Sci. 1989, 72, 2247-2256.
7.   Parris, N.; Barford, R. A.; White, A. E.; Mozersky, S. M.   J. Food Sci. 1989, 54, 1218-1221.
8.   Parris, N.; White, A. E.; Farrell, H. M., Jr.   J. Agric. Food Chem. 1990, 38, 824-829.
9.   Towbin, H.; Staehelin, T; Gordon, J.   Proc. Natl. Sci. USA 1979, 76, 4350-4354.
10. Altman, L. W.; Schneider, B. G.; Papermaster, D. S.   J. Histochemistry and Cytochemistry 1984, 32, 1217-1223.
11. Mangino, M. E.; Fritsch, D.; Liao, S. Y.; Fayerman, A. M.; Harper, W. J.   N. Z. J. Dairy Sci. Technol. 1985, 20, 103-107.
12. Pearce, K. N.; Kinsella, J. E.   J. Agric. Food Chem. 1978, 26, 716-725.
13. Phillips, L.; Haque, Z.; Kinsella, J. E.   J. Food Sci. 1987, 52, 1074-1079.

RECEIVED October 18, 1990

Chapter 4

Quantitation of Hydrophobicity for Elucidating the Structure–Activity Relationships of Food Proteins

S. Nakai, E. Li-Chan, M. Hirotsuka, M. C. Vazquez, and G. Arteaga

Department of Food Science, University of British Columbia, Vancouver,
British Columbia V6T 1W5, Canada

Hydrophobicity is a major structural variable used for
predicting the functionality of food proteins, such as
emulsification and foaming abilities. However, the
published data for protein hydrophobicity often show
inconsistencies, due to differences in the principles
underlying the various methods currently employed for
quantitation of hydrophobicity. New approaches using
nuclear magnetic resonance and Raman spectrophotometry
are proposed which may help to clarify the definition
of "surface" or "available" hydrophobicity of proteins
important for function, and should be investigated for
quantitating the extent of exposure of aliphatic and
aromatic hydrophobic side chains of protein molecules.
Once reliable and quantitative measurements of protein
hydrophobicity are obtained, these parameters may be
incorporated into QSAR equations, for use in computer-
aided optimization of food formulations.

Quantitative structure-activity relationship (QSAR) techniques use
molecular structure and physical property data to make predictions
about activity or reactivity of compounds. These techniques have
gained wide acceptance and application especially in toxicological
or pharmacological research. Various molecular structure and
property descriptors are used for the formulation of QSAR. For
example, hydrophobicity, topological descriptors, electronic
descriptors and steric effects have been suggested to predict
therapeutic response or toxicity of chemicals. Polarity has also
been a commonly used concept in the field of chemistry to explain
behavior such as solubility of compounds and the mechanism of
chromatographic behavior. The terms "polarity" and "hydrophobicity"
are commonly used as antonyms, and the term "hydrophobic" is often
synonymously used with "lipophilic" or "nonpolar".

0097–6156/91/0454–0042$06.00/0
© 1991 American Chemical Society

For small compounds, relative solubility in a nonpolar solvent versus a polar solvent is commonly used as an indication of hydrophobicity. For example, octanol/water partition coefficients are often measured as a hydrophobicity parameter. In the case of proteins, which are macromolecules, solubility has been postulated to be one of the most important factors in functionality. Therefore, polarity or hydrophobicity must also have an important role in food protein functionality. However, despite the increasing recognition of its importance in protein function, the quantitation of hydrophobicity for use as a descriptive parameter in QSAR equations is not as straightforward for proteins as it is for small molecules. The difficulty arises from the lack of sound theoretical rules to define a hydrophobicity parameter which considers the influences of steric effects of protein structure.

The objectives of this chapter are to discuss the important role of hydrophobicity in elucidating the structure-function relationships of food proteins, to review current trends and propose new approaches in quantitating protein hydrophobicity for QSAR, and to illustrate the potential benefits of applying the derived QSAR equations for optimization of food formulations.

Selection of Important Descriptors for QSAR of Food Proteins

Charge, hydrophobicity and steric parameters were originally proposed as the three major classes of descriptors for QSAR investigation for smaller compounds, as described by Hansch and Clayton (1) and Rekker (2). These parameters include measurement or calculation of Hammett constants, dipole moments, molar refractivities and ionization potentials as electronic descriptors; octanol/water partition coefficients as hydrophobicity descriptors; and Taft constants, van der Waals radii and total surface area as steric descriptors.

The importance of the three major classes of descriptors also extends to QSAR for elucidating functionality of proteins in food systems, such as emulsifying and foaming ability and stability, gelation, coagulation, film formation, water and fat binding properties. However, the traditional types of parameters mentioned above have not been used to measure these descriptors due to the complexity of food proteins in terms of co-existence of several types of proteins and changes induced upon processing. Instead, simpler alternatives have been sought which give empirical measures of parameters related to charge, hydrophobicity and steric effects. For example, solubility is recognized in the food industry as one of the most influential properties of a protein molecule which affects its other functions. In fact, solubility is a reflection of the balance of charge and hydrophobicity of the protein molecule (3, 4) which affects its interaction with the solvent (→ "soluble") and with other protein molecules (→ "insoluble"). Measurement of solubility thus indirectly provides a descriptor for QSAR which incorporates both charge and hydrophobicity effects. Other examples of parameters used to predict food protein functionality include viscosity to reflect steric effects and intermolecular interactions, and sulfhydryl and disulfide group contents to give an indication of molecular flexibility or ability for crosslink formation.

Table I shows examples of equations describing the relationships between physicochemical descriptors (solubility, hydrophobicity, viscosity, sulfhydryl or disulfide group content) and functionality (emulsifying ability, foaming capacity, thermally induced coagulation and gelation, and fat binding capacity) of some food proteins (5-8). As shown in these equations, hydrophobicity is an important parameter to explain diverse functional properties of food proteins. Various parameters have been used to represent hydrophobicity in these equations (ANS, CPA, $CPAS_e$, and $H\phi_{avg}$) and different parameters have been used by other workers in the area for correlation to functionality (9). In most cases, these parameters measure hydrophobicity of protein molecules in dilute solutions. Whether or not the measurement of hydrophobicity of proteins at concentrations typically encountered in food applications would improve the accuracy of equations for QSAR elucidation is not clear. However, comparison of the equations for predicting coagulability and gel strength of ovalbumin solutions at 0.5 and 5.0% concentration, respectively, shows that the relative importance of various physicochemical parameters to describe functionality is concentration dependent. ANS hydrophobicity and charge frequency expressed as zeta potential (ZP) were found to be important in predicting coagulability upon heating of 0.5% ovalbumin solutions, whereas ANS hydrophobicity and sulfhydryl (SH) group content were significant in explaining gel strength of 5% heated solutions (8). Similarly, emulsifying activity of soluble proteins was dependent on the concentration of the protein solution, as shown in Figure 1 (10). At protein concentrations approaching zero, emulsifying activity expressed as turbidity of the emulsion, A_{500}, was correlated with CPA hydrophobicity of the proteins:

$$A_{500} = 0.387 + 8.391 \ CPA \qquad\qquad (r=0.932, \ P<0.05).$$

However, at protein concentrations of 2.5% or greater, a parameter describing the complex interaction of CPA hydrophobicity with viscosity (η_0) of the continuous phase became a significant factor in emulsion formation:

$$A_{500} = 0.382 + 16.52 \ \eta_0^2 CPA^{\frac{1}{2}} \qquad\qquad (r=0.922, \ P<0.05).$$

For application of QSAR equations to explain and predict functionality of food proteins, the quantitative measurement of relevant physicochemical parameters is crucial. Of the various physicochemical parameters identified to be important in explaining food protein functionality, methods have been well established to measure most of them, such as solubility, viscosity and sulfhydryl or disulfide groups. However, consensus has not yet been reached on a method for measuring hydrophobicity of a protein which can explain functionality.

Comparison of Current Methods for Measuring Protein Hydrophobicity

Methods proposed for quantitative estimation of protein hydrophobicity can be roughly categorized into (1) calculated values using data of hydrophobicity scales of the individual amino acids

Table I. Examples of regression equations reported in the literature describing the relationships between physicochemical descriptors and functionality of some food proteins

Functional Property	Regression Equation	Reference
Emulsifying Activity Index (EAI) of native and heated proteins	$EAI = 16.9 + 0.21CPA + 0.93s - 0.007s^2$ (n=52, $R^2=0.583$, P<0.001)	Voutsinas et al. (5)
Emulsion Stability Index (ESI) of native and heated proteins	$ESI = -69.5 + 0.565CPA + 2.03s - 0.004CPA\ s - 0.012s^2$ (n=49, $R^2=0.584$, P<0.001)	Voutsinas et al. (5)
Fat Binding Capacity (FBC) of native and heated proteins	$FBC = 4.90 + 0.45CPA + 1.40s - 0.001CPA^2 - 0.014s^2$ (n=48, $R^2=0.473$, P<0.001)	Voutsinas et al. (5)
Foaming Capacity (FC) of native proteins	$\ln(FC +30) = 0.039 + 0.0041H\emptyset_{avg}$ (n=11, $r^2=0.677$, P<0.01) or $FC = -1775 + 0.1493\eta + 25.93\ \ln Se$ (n=19, $R^2=0.779$, P<0.01)	Townsend & Nakai (6)
Water Absorption (AM_b) of minced meat in brine	$AM_b = -0.26 + 0.0021CPA - 0.0000017CPA^2$ (n=58, $R^2=0.439$, P<0.001)	Li-Chan et al. (7)
Coagulability (C) of native and heated 0.5% ovalbumin solutions	$C = -4.77 + 0.476ANS - 0.000404ANS^2 - 0.0137ANS\ ZP$ (n=26, $R^2=0.794$, P<0.001)	Hayakawa & Nakai (8)
Gel Strength (G) of native and heated 5.0% ovalbumin solutions	$G = 821 - 0.0628ANS - 8.91SH$ (n=26, $R^2=0.621$, P<0.001)	Hayakawa & Nakai (8)

Abbreviations:
ANS = hydrophobicity determined using 1-anilinonaphthalene-8-sulfonic acid.
CPA = hydrophobicity determined using cis-parinaric acid.
CPASe = hydrophobicity determined using cis-parinaric acid after first treating the protein solution by heating in the presence of sodium dodecyl sulfate (6).
$H\emptyset_{avg}$ = average hydrophobicity value calculated according to the method of Bigelow (3).
η = viscosity.
SH = sulfhydryl group content.
ZP = net charge determined as zeta potential.

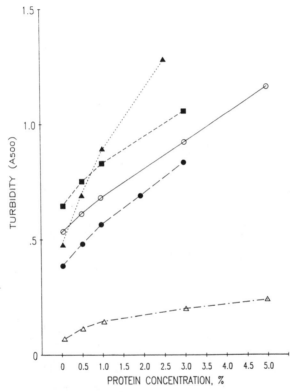

Figure 1. Effect of protein concentration on emulsifying activity.
○ , bovine serum albumin; △ , ovalbumin; ● , casein; ▲ , gelatin;
■ , β-lactoglobulin.

and the amino acid composition of the protein; (2) partition methods, including relative solubility in polar and nonpolar solvents, or relative retention on reverse phase or hydrophobic interaction chromatography; (3) binding methods, including the binding of aliphatic and aromatic hydrocarbons, sodium dodecyl sulfate, simple triglycerides and corn oil; (4) contact angle measurement; and (5) spectroscopic methods, including intrinsic fluorescence, derivative spectroscopy, and use of fluorescence probes. These methods have been recently reviewed (9, 11), and some of the published data for protein hydrophobicity using different methods are compared in Table II.

Considerable variations are observed for the values obtained by different methods, as shown in Table II. Some inconsistencies in these values could be attributed to the fact that the sample proteins used for analyses may not have been from the same source, or that the conditions such as pH, ionic strength or concentration were not necessarily the same. Nevertheless, dramatic differences are apparent. For example, large values for hydrophobicity were reported for both chicken albumin and lysozyme when measured by hydrophobic interaction chromatography and for chicken albumin by calculation of average hydrophobicity, compared to the small values obtained by the fluorescence probe, binding and partition methods. It is probable that differences in principles underlying the various methods of measuring hydrophobicity are the cause of the significant variation in the reported values.

Even for methods within a class, such as reverse phase (RP) and hydrophobic interaction chromatography (HIC), discrepancies have been observed. For example, there was no correlation between the retention times of 12 proteins eluted by RP and HIC (12). Recently, anomalous elution behavior of proteins on HIC columns was explained by salt binding, which may alter the number and distribution of protein surface groups, including charged groups (13). Steadman et al. (14) stated that it is important to distinguish between those methods which estimate hydrophobicity by measuring aggregation from those relying on differential solubility as well as those estimating the adherence of substances. According to the Gibbs adsorption isotherm, the adsorption of solutes is negatively correlated to the difference in surface tension between solute and solvent, and the polarity of the solute seriously affects this difference. It is reasonable therefore to assume that hydrophobic attraction plays a major role in adsorption chromatography.

Wilson et al. (15) investigated RP chromatographic behavior of 96 peptides, ranging in length from 2 to 65 residues. Hydrophobic constants of amino acid residues were computed from the retention properties of these peptides, and these constants were compared with hydrophobic constants published in the literature obtained by other methods. Correlation coefficients between these constants and peptide retention times were also computed. Table III shows that the best correlation was obtained for the constants derived by RP chromatography by Wilson et al. (15). It was thus concluded that chromatographic behavior of peptides could be explained based on polarity of the constituent amino acid residues.

Although chromatographic behavior of peptides could be successfully predicted, the situation is not as straightforward for

Table II. Relative hydrophobicity values of some proteins measured by different methods [a] (Reproduced with permission from Ref. 11. Copyright 1990 Wiley-Interscience.)

Proteins	HØ$_{ave}$	fluorescence probes			DPH probe TG binding	chromatography		binding			partition Δlog K
		S_0ANS	S_0CPA	S_eCPA		RPC	HIC	heptane	SDS	TG	
albumin, bovine	1120	1000	100	100	100	100	100	100	100	100	100
albumin, chicken	1110	7	2	96	15	0	9-86	2	14	3	1-25
casein, α-	1200	57	6-30	107	150	400	/	/	/	160	/
casein, β-	1320	60	50	/	/	315	/	/	/	/	/
casein, κ-	1210	83	13-93	89	/	340	/	/	30	/	/
chymotrypsin, α-	1030	0	3	/	41	/	57-139	9	/	3	9-33
globulin, soy 7S	1090	47	9-77	/	50	165	/	/	21	/	/
globulin, soy 11S	950	27	2-17	/	25	150	/	/	/	/	/
lactalbumin, α-	1150	33	9-54	/	98	225	10	/	/	/	/
lactoglobulin, β-	1230	13	54-146	80	49	0	67-102	100	62	130	41-78
lysozyme	970	0.7	7	7	/	/	42-113	0	19	1	0
ovomucoid, chicken	920	0	0	22	/	/	/	24	18	108	/
pepsin	1063	0.7	2	57	32	/	92	17	18	95	/
ribonuclease A	870	/	1	24	/	/	4-71	29	20	2	9-20
transferrin, chicken	1080	/	4	108	/	/	16-31	/	13	118	9-27
trypsin	940	8	3	34	/	/	/	/	/	/	/
trypsin inhibitor	1040	0	0	/	1	0	/	/	/	/	/

[a] Adapted from ref 11. The majority of the data are expressed relative to "100" for bovine albumin to facilitate comparison. Where varying values were reported by investigators using essentially the same method, a range of values is presented.

[b] 530 and 800 for α- and β-casein, respectively, at the isoelectric pH.

Abbreviations: HØ$_{ave}$=average hydrophobicity calculated by Bigelow's method (3); S_0=initial slope of relative fluorescence intensity versus protein concentration plot, using native proteins; S=S_0 measured for protein solutions after treatment with 1.5% SDS at 100°C for 10 min; ANS=1-anilinonaphthalene-8-sulfonic acid; CPA=cis-parinaric acid; DPH=1,6-diphenyl-1,3,5-hexatriene; TG=triglyceride; RPC=reverse phase chromatography; HIC=hydrophobic interaction chromatography; SDS=sodium dodecyl sulfate; /=no data available.

Table III. Amino acid hydrophobicity constants and correlation with peptide chromatographic retention times (Adapted from Ref. 9. Reproduced with permission. Copyright 1988 CRC Press.)

amino acid	Bigelow-Chapman	Meek	Pliska-Fauchere	Rekker	Segrest-Feldman	Wilson et al.
			Amino Acid Hydrophobicity Constants			
Ala	0.5	-0.1	0.38	0.53	1.0	-0.3
Arg	0.75	-4.5	-1.23	-0.82	-	-1.1
Asn	-	-1.6	-0.27	-1.05	-1.5	-0.2
Asp	0.0	-2.8	-1.23	-0.02	-	-1.4
Cys	-	-2.2	-	1.11	0.0	6.3
Gln	-	-2.5	-0.09	-1.09	-1.0	-0.2
Glu	0.0	-7.5	-1.20	-0.07	-	0.0
Gly	0.0	-0.5	0.0	0.0	0.0	1.2
His	0.50	0.8	-1.3	-0.23	1.0	-1.3
Ile	2.95	11.8	1.56	1.99	5.0	4.3
Leu	1.80	10.0	1.66	1.99	3.5	6.6
Lys	1.50	-3.2	-0.93	-0.52	-	-3.6
Met	1.30	7.1	1.39	1.08	2.5	2.5
Phe	2.50	13.9	1.80	2.24	5.0	7.5
Pro	2.60	8.0	0.56	1.01	1.5	2.2
Ser	-0.30	-3.7	0.04	-0.56	-0.5	-0.6
Thr	0.40	1.5	-0.33	-0.26	0.5	-2.2
Trp	3.40	18.1	1.87	2.31	6.5	7.9
Tyr	2.30	8.2	1.70	1.70	4.5	7.1
Val	1.50	3.3	1.06	1.46	3.0	5.9
Correlation coefficient[a]						
(1) with RT	0.536	0.681	0.713	0.693	0.826	0.831
(2) with 15	0.741	0.757	0.868	0.841	0.778	-

[a] Correlation coefficients were computed (1) between experimental and predicted retention times RT of peptides, calculated using the amino acid hydrophobicity constants in each column; and (2) between amino acid hydrophobicity constants in each column compared to those of Wilson et al. (15).

the behavior of amphiphilic macromolecules such as proteins. In the latter case, the retention times cannot usually be accurately determined based on hydrophobicity values calculated by simply summing up hydrophobic constants of the constituent moieties of the macromolecules, due to the possible existence of steric effects. The problem is further compounded by the influence of highly nonpolar or organic solvents often used in reverse phase chromatography or partition methods, which can destroy the so-called native structure of proteins, thus changing the steric effects.

A significant limitation of values calculated from hydrophobicity scales of amino acids is their lack of consideration of the effect of tertiary and quaternary structures of proteins on the extent of exposure or "effective" hydrophobicity of residues in individual proteins. It is generally agreed by protein chemists that charged residues are located preferentially at the surface of the molecule, where they can interact with water; residues in the interior are close packed and burial of hydrophobic groups away from the surface can be a major source of stabilization of tertiary structure. It is therefore likely that the groups which can participate in protein functionality are those hydrophobic residues which are located on the surface of the native protein molecules or become exposed during processing such as heating or whipping - in other words, "surface" or "available" hydrophobicity.

Methods such as the fluorescence probe or the various ligand binding techniques measure hydrophobic groups on the surface of the protein molecule which are able to bind the probes or ligands, and thus are expected to yield parameters which correlate with the functionality of proteins. In terms of simplicity of methodology, hydrophobic probe methods using anilinonaphthalenesulfonate (ANS), cis-parinaric acid (CPA) and other fluorescence probes are probably the most popular for hydrophobicity determination. Good correlation has been observed between the surface hydrophobicity of proteins measured by these probes and functionality. However, criticism has arisen against using ANS and CPA as strictly hydrophobic probes because of the coexistence of charge bearing moieties on these probe molecules which can interact with the protein. Considering this, diphenyl hexatriene (DPH) may be a more suitable probe as it is nonpolar and nondissociable. To overcome the insolubility of DPH in water, Tsutsui et al. (16) first dissolved DPH in corn oil, then measured the fluorescence associated with oil bound to proteins. Although strictly speaking, this method measures the oil-binding capacity rather than hydrophobicity of proteins, it may yield a useful physicochemical parameter in elucidating food protein functionality because of the importance of lipid- or oil-protein interactions in food systems, eg. emulsifying and fat binding properties. However, despite its potential relevance for explaining food protein functionality, this method has not been widely used probably due to the rather tedious and time-consuming nature of the procedure.

As shown in Table I, various hydrophobicity parameters have been used to develop equations explaining functionality. Although the fluorescence probe methods with ANS and CPA have proved most popular due to ease of measurement and ability to predict functionality, application of the hydrophobicity values obtained by these methods may be limited due to (1) the presence of the anionic group on these probes which may interact with the protein molecule through charge effects, and (2) the low protein concentrations used for measurement of fluorescence, which is in contrast to the much higher concentrations usually encountered in real food systems.

New Approaches to Determination of Protein Hydrophobicity

Since Wilson et al. (15) have shown that the hydrophobic constants of individual amino acids can be successfully used to predict behavior such as retention on reverse phase chromatography for peptides in which steric effects are negligible, one approach to determination of protein hydrophobicity would be to quantitate the extent of exposure of different types of amino acid residues in the protein and then use established amino acid hydrophobicity scales to calculate the total hydrophobicity of the exposed residues. Two methods are currently being investigated in our laboratory for their ability to quantitate the extent of exposure of side chains in proteins, namely proton magnetic resonance (PMR) and laser Raman spectroscopy.

Proton Magnetic Resonance Spectroscopy. According to McDonald and Phillips (17) who developed a procedure to compute PMR spectra of random coil proteins from their amino acid composition, protein molecules in 6M guanidine at 40°C appear to be in a configuration close to random coil; all hydrocarbon side chains then appear at 1-1.7 ppm and aromatic at 6-8 ppm. Wurthrich and Wagner (18) also stated that the spatial folding of the polypeptide chain is responsible for the differences between PMR spectrum observed for a native globular protein and the spectrum observed for the random coil form of the polypeptide chain, which corresponds closely to the computed spectrum.

Kason et al. (19) in their interaction study of α_{s1}-casein showed sharpening effects of pH and concentration of the protein on its PMR spectra (Figure 2). The manually computed PMR spectrum (Figure 3) showed good agreement with that of the most dissociated spectra at 2.5% protein concentration and pH 9.9. A BASIC program of the procedure of McDonald and Phillips (17) was written for a PC computer as shown in the same figure for comparison.

The differences in the measured spectrum of a protein solution from the theoretical or computed spectrum are due to the deviation from a completely random coil structure in the protein solution, i.e. not all of the amino acid residues are exposed. Computation of a spectrum which matches the real spectrum thus should give information of the number of exposed amino acids in the protein. The program for PMR spectrum computation was incorporated into a computer program for simplex optimization to find the amino acid composition yielding a computed spectrum most closely resembling the measured protein spectrum. The differences in amino acid compositions computed to match spectra of the protein solution measured under various conditions such as pH, temperature or denaturing agent should correspond to differences in extent of exposure of the amino acid residues of the protein molecules. The optimization program being used in this study is a slight

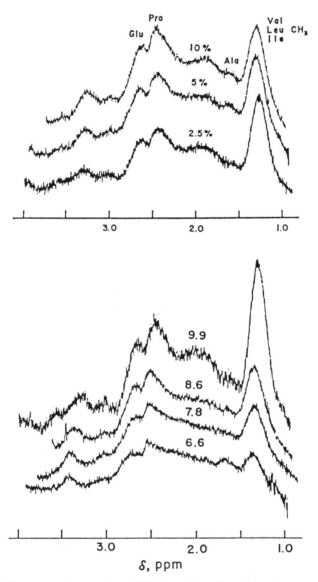

Figure 2. PMR spectra of α_{s1}-casein at 100 Mc/sec. Top, effect of protein concentration at pH 10; bottom, effect of pH. (Reprinted with permission from Ref. 19. Copyright 1971 American Dairy Science Association.)

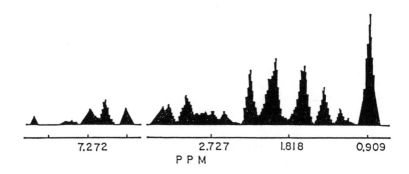

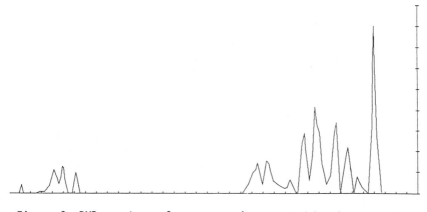

Figure 3. PMR spectrum of α_{s1}-casein computed by the procedure of McDonald and Phillips (17). Top, manually computed; bottom, computer-drawn.

modification of the program we have used for determining the
blending ratio of ingredients to yield a GC pattern for the blended
sample most similar to that of a reference standard, by maximizing
a pattern similarity constant (20). The current optimization of PMR
spectra is being carried out separately for aromatic and aliphatic
regions of the spectrum. This approach has been validated using
model PMR spectra, as the computed spectra satisfactorily matched
with the original model spectra, and application to measured PMR
spectra of a variety of proteins is underway.

Raman Spectroscopy.   Raman spectroscopy has been used for
determining the microenvironment of aromatic side chains in protein
molecules (21,22). The intensity ratio I_{850}/I_{830} is used for
estimating the exposure of tyrosine side-chains to water, and the
appearance of peaks at 760, 880 and 1360 cm^{-1} are used as a sign
for tryptophan side-chains in hydrophobic environment. Changes in
C-C and C-H stretch bands at 950 and 1449 cm^{-1} were observed by
thermal denaturation of egg white (23) and could represent indirect
evidence of a change in the microenvironment of hydrocarbon side
chains. Increase in the CH_3-stretching region at around
2930 cm^{-1} was suggested to arise from the insertion of previously
buried aliphatic side chains into water (24). A significant
difference in the application of Raman compared to other forms of
spectroscopy is its ability to measure these changes in protein
solutions of high concentration, or in solid or gelled states.
     Use of Raman spectroscopy in the study of hydrophobic
interaction is exemplified in our recent work (25). A Jasco model
NR-1100 equipped with a 488 nm Argon laser was used for
investigating interaction between 7S soy protein and soy lecithin.
The difference spectrum between 7S-lecithin complex and their
noninteracting mixture is shown in Figure 4, and characteristic peak
assignments are shown in Table IV. The difference spectrum indicates
decreases in intensity of peaks corresponding to α-helix and
vibration of C-N stretch in 7S protein, and to C-C and hydrocarbon
vibration in the phosphatidylethanolamine component, while increases
in intensity are observed for peaks corresponding to β-sheet in 7S
and to hydrocarbon vibration in phosphatidylcholine. Exposure of
hydrophobic amino acid side chains thus observed was in agreement
with the increase in hydrophobicity analyzed fluorometrically using
CPA as a probe.

Application of QSAR to Formula Optimization

Once the structure-function relationships have been established, the
best area for their use may be found in food formulation.
     Linear programming (LP) is the most popular computer-aided
technology currently being used for food formulation. For example,
the processed meat industry uses LP in determination of the least
cost formula that will meet predetermined product specifications
using available ingredients. The specifications that are used as
constraints in the LP computer program include proximate
composition, ingredient content, and quality in the form of bind
constants (26). However, the least cost programs place excessive
emphasis on cost reduction and unduly deemphasize the product

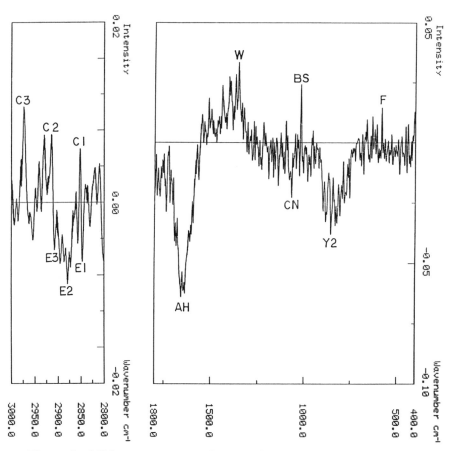

Figure 4. Difference spectrum between 7S protein-lecithin complex and noninteracted mixture. The symbols are explained in Table IV.

Table IV. Peak characteristics of difference spectrum between
7S-lecithin complex and noninteracted mixture shown in Figure 4

Symbol	Frequency (cm^{-1})	Assignment
F	624	Phe
Y1	830	Tyr
Y2	850	Tyr
BS	1002	Beta-sheet
CN	1110	C-N
W	1361	Trp
AH	1650	Alpha-helix
E1	2848	CH_2 in PE^a
C1	2850	CH_2 in PC^b
E2	2855	CH_2 in PE
E3	2900	CH_3 in PE
C2	2920	CH_2 in PC
C3	2960	CH_3 in PC

[a] PE = phosphatidylethanolamine.
[b] PC = phosphatidylcholine.

quality, by dealing with quality parameters as constraints rather than objective functions. The bind constants frequently used as quality parameters by the meat processors are relative variables, and linear relationships with the ingredients are assumed. However, the structure-functionality relationships of meat proteins are described by non-linear functions (7). Furthermore, relationships between ingredient composition and final product quality have also been shown to be nonlinear, making the incorporation of the prediction equations into the LP program difficult. It was found that the constrained simplex optimization (Complex) of Box (27) is more appropriate to use for this kind of formula optimization (28). Prediction equations are required in this formula optimization to define the objective functions as well as any imposed constraints. QSAR equations can best be used for this purpose. However, as discussed above, QSAR for food protein functionality are currently limited by the lag in progress to obtain an accurate or functionally relevant quantitative parameter to describe hydrophobicity in these equations.

Advances in this area will be a major step towards achieving the long term goal of elucidation and prediction of the functional properties of ingredients by QSAR analysis to ensure high standards of quality in food products.

Conclusions

The establishment of equations which can describe the quantitative structure-activity relationship (QSAR) of food proteins depends on accurate quantitation of hydrophobicity values for use as a descriptive parameter to elucidate functionality. New approaches to measure extent of exposure of amino acid residues, using computer-aided curve fitting of PMR spectra or monitoring changes in the intensity of peaks in Raman spectra, are proposed to complement current methods of hydrophobicity measurement. Once quantitative structure-functional property relationships are known, they can be used for optimization of food formulations.

Acknowledgments

The authors are indebted to Mrs. N. Helbig and L. Kwan for their technical assistance. This work was supported by an operating grant from the Natural Science and Engineering Council of Canada.

Literature Cited

1. Hansch, C.; Clayton, J. M. J. Pharm. Sci. 1973, 62, 1-21.
2. Rekker, R. F. The Hydrophobic Fragmental Constant; Elsevier: New York, 1977.
3. Bigelow, C.C. J. Theoret. Biol. 1967, 16, 187-211.
4. Hayakawa, S.; Nakai, S. J. Food Sci. 1985, 50, 486-491.
5. Voutsinas, L. P.; Cheung, E.; Nakai, S. J. Food Sci. 1983, 48, 26-32.
6. Townsend, A.-A.; Nakai, S. J. Food Sci. 1983, 48, 588-594.
7. Li-Chan, E.; Nakai, S.; Wood, D. F. J. Food Sci. 1987, 52, 31-41.

8. Hakayawa, S.; Nakai, S. Can. Inst. Food Sci. Technol. J. 1985, 18, 290-295.
9. Nakai, S.; Li-Chan, E. Hydrophobic Interactions in Food Systems; CRC Press: Boca Raton, Florida, 1988.
10. Nakai, S.; Li-Chan, E. 195th ACS National Meeting. 1988.
11. Li-Chan, E. In Encyclopaedia of Food Science and Technology; Hui, Y.H., Ed.; Wiley-Interscience: New York, NY, 1990, in press.
12. Fausnaugh, J. L.; Kennedy, L. A.; Regnier, F. E. J. Chromatog. 1984, 317, 141-155.
13. Barford, R. A.; Kumosinski, T. F.; Parris, N.; White, A. E. J. Chromatog. 1988, 458, 57-66.
14. Steadman, R.; Topley, N.; Knowlden, J. M.; MacKenzie, R. K.; Williams, J. D. Biochim. Biophys. Acta 1989, 1013, 21-28.
15. Wilson, K. J.; Honegger, A.; Stotzel, R. P.; Hughes, G. J. Biochem. J. 1981, 199, 31-41.
16. Tsutsui, T.; Li-Chan, E.; Nakai, S. J. Food Sci. 1986, 51, 1268-1272.
17. McDonald, C. C.; Phillips, W. D. J. Amer. Chem. Soc. 1969, 91, 1513-1521.
18. Wurthrich, K.; Wagner, G. Trend Biochem. Sci. 1978, 227-230.
19. Kason, W. R; Nakai, S.; Bose, R. J. J. Dairy Sci. 1971, 54, 461-466.
20. Aishima, T.; Wilson, D. L.; Nakai, S. In Flavour Science and Technology; Martens, M.; Dalen, G. A.; Russwurm, H. Jr., Eds.,; John Wiley & Sons: Chichester, 1987; p 501.
21. Carey, P. R.; Fast, P.; Kaplan, H.; Pozsgay, M. Biochim. Biophys. Acta 1986, 872, 169-176.
22. Hamodrakas, S. J.; Kamitos, E. I.; Papadopoulou, P. G. Biochim. Biophys. Acta 1987, 913, 163-169.
23. Painter, P. C.; Koenig, J. L. Biopolymers 1976, 15, 2155-2166.
24. Verma, S. P.; Wallach, F. H. Biochem. Biophys. Res. Comm. 1977, 74, 473-479.
25. Hirotsuka, M.; Nakai, S. Ann. Meeting, Inst. Food Technol. 1990.
26. Pearson, A. M.; Tauber, F. W. Processed Meats, AVI: Westport, CT, 1984.
27. Box, M. J. Computer J. 1965, 8, 42-52.
28. Vazquez-Arteaga, M. C.; Nakai, S. Ann. Meeting, Inst. Food Technol. 1990.

RECEIVED June 20, 1990

Chapter 5

Milk Protein Ingredients

Their Role in Food Systems

Steve J. Haylock and Wayne B. Sanderson

New Zealand Dairy Research Institute, Private Bag,
Palmerston North, New Zealand

The roles which milk protein ingredients play in food systems are diverse. In multi-component food products, interactions occur between ingredients. The characteristics of these food products depend considerably on interactions. Predicting changes in the characteristics of food products from the physico-chemical properties of ingredients is difficult at present. The results of interactions between ingredients can be evaluated by measuring the changes in food product characteristics, caused by changes in the levels of ingredients. Examples showing the application of milk protein ingredients in a water-added ham and a yellow layer cake are discussed. In both cases desirable product characteristics have been optimized through the interaction of milk protein ingredients with other components in the food system.

Milk protein ingredients are used by many sectors of the food industry (1-3). These ingredients are produced with a number of different properties and they are used in food systems for a wide range of purposes.

Studies on the functional properties of isolated milk protein products continue to be carried out and the results from these illustrate the versatility of milk proteins in the range of physico-chemical or functional properties which they are able to demonstrate (4-5).

Rather less information though, has appeared in the literature on the roles which milk proteins assume in food systems. Some of the reasons for this are the confidential nature of many food formulations but equally responsible is the complex nature of many of todays formulated foods and the difficulty in interpreting the functional contribution of individual ingredients.

0097–6156/91/0454–0059$06.00/0
© 1991 American Chemical Society

Milk Proteins

There are two fundamentally different groups of proteins present in milk, the casein proteins and the whey proteins. It is possible to distinguish between these two groups of proteins on the basis of the properties they exhibit. A number of these properties are shown in Table I (3).

Table I.   Properties of milk proteins

Protein Type	Properties
Casein	Contains strongly hydrophobic regions Contains little cysteine Random coil structure Heat stable Unstable in acidic conditions
Whey	Balance of hydrophilic and hydrophobic residues Contains cysteine and cystine Globular structure, much helical content Easily heat denatured Stable in mildly acidic conditions

Milk protein manufacturers use these different characteristics as the basis for isolating or separating different protein components from milk.   Casein proteins are isolated from skimmed milk using precipitation.   This can be brought about by acidification of the milk to pH 4.6, or, by encouraging the hydrophobic interaction of casein molecules by removing the hydrophilic part of the molecule with the action of rennet enzyme.

Whey proteins are far less sensitive to acidic pH conditions and consequently after removal of the casein proteins, these can be purified from the remaining carbohydrate and mineral material by filtration, ion exchange or heat-precipitation techniques.

The protein products which are isolated from casein and whey consists of a number of individual proteins.   Table II shows the major proteins in each group and their percentage contribution to the total protein in milk (3).

The isolation of individual casein and whey proteins is possible in the laboratory and in some instances in the pilot plant and manufacturing plant (6-7).   A potentially rewarding area of research will be the isolation of usable quantities of individual proteins for evaluation of their functional properties and their performance in food systems.

Ingredient Manufacture and Usage

The isolation and separation techniques used for casein and whey proteins form the basis of designing specific milk protein ingredients.   The manufacturers of these ingredients though, use a number of other techniques along with these isolation procedures in creating dairy ingredients with the required properties. Some of the more important techniques used by manufacturers to modify the properties of the milk proteins are shown in Table III (8).

Table II.   Major protein components in milk

Protein Type	Individual Proteins	Total Milk Protein (%)
Casein	α_s casein	45–55
	β casein	23–35
	K casein	8–15
	casein	3–7
Whey	β lactoglobulin	7–12
	α lactalbumin	2–5
	Proteose peptone	2–6
	Immunoglobulins	2–3
	Bovine Serum Albumin	ca 1

Much of the information detailing how these techniques are able to affect the properties of milk protein ingredients is proprietary to ingredient manufacturers. Some information has been published though, on pH/temperature/time treatments and the adjustment of mineral balance (9–12). One or a number of these techniques may be selected in order to enhance the properties of the ingredients required.

The reasons why dairy ingredients based on milk proteins are used so extensively by food manufacturers lie with the varied range of functional properties which they exhibit, their excellent nutritional quality and their overall ability to contribute beneficially to formulated food systems. Much has been written about the functional properties exhibited by milk proteins and a number of these which have been discussed in the literature are listed in Table IV (4,5,13–15).

Many of the functional properties which have been discussed in the scientific literature relate to properties exhibited by those proteins in simple solutions rather than in complex, multi-component systems which are typical of todays formulated foods. There is no doubt that dairy proteins are extremely useful ingredients to the food manufacturer. There is also no doubt that these proteins exhibit a broad range of functional properties in simple systems. What is not clear however, is the relationship between these simple functional properties and the role of proteins, as ingredients, in complex food systems. Experiences in our laboratories confirm findings in the literature which show that as soon as other components are added to proteins in simple test situations then the original functional properties of the protein become considerably modified (16).

This creates a problem for the dairy ingredient chemist in not being able to relate functional properties to a required response in a food system. However, on the other hand the fact that other ingredients in food systems are capable of modifying the functional properties of milk proteins means that there is an additional challenge to food technologists. That is, to understand which significant interactions will occur between ingredients and how to exploit these to bring about desirable properties in food systems.

Table III.   Protein Modification Techniques Used in
Ingredient Manufacture

Adjustment of mineral balance (cations and anions)
Hydrolysis
Temperature/time treatments
pH/time treatments

Selection of protein concentration
Use of complementary ingredients

Table IV.   Functional Properties Exhibited by Milk Proteins

Functional Property	Protein Type
Foam formation and stability	Casein and Whey
Emulsification and fat binding	Casein and Whey
Gelation	Whey
Heat stability	Casein
Water binding	Casein and Whey
Viscosity and texture modification	Casein
Solubility - neutral pH	Whey and casein
- acidic pH	Whey
Solution opacity	Casein

Food Systems Using Milk Protein Ingredients

Although there is difficulty in determining the relationship between simple solution functionality and complex food system functionality, nevertheless many different protein products are used in a wide range of foods.   Table V shows a number of the more significant food applications where milk protein ingredients are used (17-23).

It is likely that in all the food products shown in Table V, interactions will occur between the milk proteins and other components in the formulation, particularly where the milk protein is used for a functional purpose as compared to a nutritional purpose.

Given the nature of the molecular structures of the milk proteins it is possible to envisage a number of different mechanisms for interactions between milk proteins and other food ingredients. These include hydrophobic interactions, hydrophilic interactions, salt bridges, hydrogen bonding and sulphydryl reactions (15).

Table V.   Food Products Utilizing Milk Protein Ingredients

Category	Product Type	Protein Used
Analogue dairy products	Whipped topping	Casein
	Coffee whitener	Casein
	Imitation cheese	Casein
	Margarine	Casein
	Low fat spreads	Casein
Bakery	Egg replacers – cakes & pastry	Whey
Beverages	Acidic beverages	Whey
Confectionery	Marshmallow	Casein
	Caramel	Casein, Whey
Meat	Sausage products	Casein
	Ham products	Whey
	Fish products	Whey, Casein
	Poultry products	Casein, Whey
Nutritional products	Diet formula	Casein, Whey
	Infant formula	Whey, Casein
Dairy products	Processed cheese	Casein
	Yoghurt	Casein, Whey

Interactions in Food Systems

Formulated foods can contain a wide variety of ingredients: fats, proteins, salts, stabilizers, emulsifiers, acidulants, simple sugars etc. In assessing the likelihood of potential interactions it must be assumed that if an ingredient is present then it is likely to be able to interact with the other ingredients in the food formulation (6). Making predictions then, about the properties of a food system on the basis of interactions which are likely to occur is a difficult way to determine the value of ingredients to a particular system. The results of interactions with protein ingredients in food systems, though, can be measured. This can be done by measuring the response of particular food system parameters to changes in the concentration of the protein ingredients. Two food systems which have been studied in this way are the water-added ham and the yellow layer cake.

Ham Products. These products are popular in many parts of the world. They range from traditional dry cured bone in ham products, through to products where the raw meat can form only 50% of the finished article. Intermediate within this range are a number of cured pork products where the pork meat is only 60-75% of the cooked product.

At lower extension levels the extraction of myofibrillar meat proteins using salt and phosphate is generally sufficient to provide the gelling and waterbinding properties required in ham products. Higher levels of extension though are often the norm in many countries and as a consequence of this, a range of food ingredients are permitted in these products.   Both caseinate and whey protein ingredients are used in these systems for their waterbinding and gelling properties. These ingredients are able to interact with and complement the soluble myofibrillar meat proteins.

A series of water-added ham products were prepared in order to optimize the ratio of two milk protein ingredients, Whey Protein Concentrate (WPC(x)) and Total Milk Proteinate (TMP(y)).   A formulation which gave a 60% extension of the original pork meat was used. The steps involved in the manufacturing process for these hams are shown in Table VI.

Food products can never be characterized by a single property and water-added hams are no exception to this. There will always be a balance between a number of properties which will combine together to make an acceptable product. The factors which are considered to be of major importance in a water-added ham product are shown in Table VII along with the relative weighting of these factors.

The concentration of WPC(x) was varied between 1.0% and 2.5% while the concentration of TMP(y) was varied between 0.5% and 1.5%. So that the concentration of the raw meat and brine was not varied, lactose was used to balance the formulation. The combined addition level for WPC(x), TMP(y) and lactose was 3.5%.

Table VI.   Manufacturing Process for Water-Added Ham Products

Raw Meat Preparation	Milk Protein Hydration	Brine Preparation
(fat trimming, tenderising grinding)	(water, WPC(x), TMP(y)	(water, ice, salt, dextrose, sodium tripolyphosphate, sodium hexametaphosphate, sodium erythrobate, sodium nitrite)

Tumbling under Vacuum

Forming Into Impermeable Casings

Cooking or Smoking

Table VII.   Product Characteristics for Water-Added Ham Products

Product Characteristics	Percentage Weighting
Cooked yield	35
Syneresis	20
Texture - Breakstrength	15
- Rigidity	12.5
- Elasticity	17.5

     Cooked yield was determined by dividing the weight of the cooked ham by the weight of the raw ham. Syneresis was determined by measuring the moisture loss after storage of sliced product for 14 days in vacuum packs. Texture measurements were carried out using an Instron Universal testing machine. The break-strength was defined as the pressure required to break a 60 mm slice of ham. The rigidity was defined as the pressure required to deform the ham by 8 mm. The elasticity was defined as the distance the ham was compressed prior to breakage, divided by the thickness of the slice.

     Figures 1(a - c) and 2(a and b) show response surface diagrams of the product properties which were measured and how these were affected by the changing levels of the two protein ingredients. It is apparent from these figures that the optimization of a formulation for a certain product property will often be to the detriment of other properties. Under these circumstances a compromise must be reached.

     The relative importance or the weighting assigned to the different product properties may be varied depending on the product required. The weighting factors are generally assigned subjectively by an experienced worker in the area.

     Figure 2c shows a further response surface diagram. In this case the z axis is the compromise metric which takes into account each of the measured properties. The best compromise formulation with regard to the levels of WPC(x) and TMP(y) is when z is at a minimum value.

Egg Replacer.   Whole eggs are regarded as multi-functional ingredients in a number of food systems in which they are used (24). This is because they consist of a complex mixture of proteins, lipids and phospholipids (25).   It is not surprising that ingredient manufacturers who have attempted to imitate the functionability of egg products, by supplying single component products as egg substitutes have met with, at best, moderate success to date.

     While the development of a universal egg substitute is perhaps some way off, it is possible to imitate the functionality of egg, to a substantial degree, in some specific food applications.

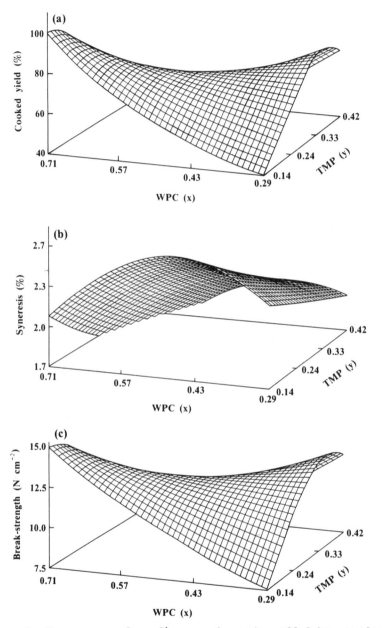

Figure 1. Response surface diagrams for water-added ham products.
(a) Cooked yield, (b) Syneresis, (c) Break-strength.

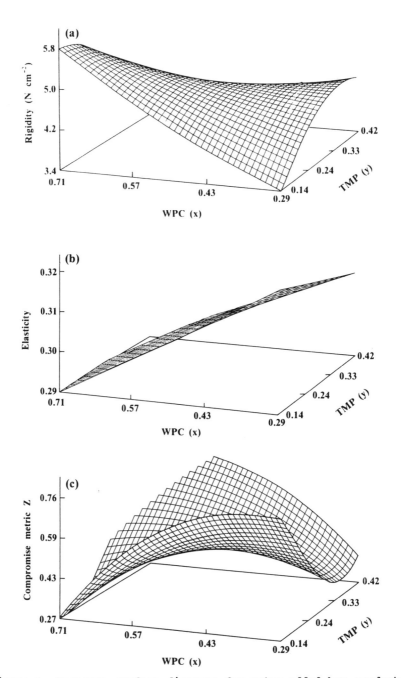

Figure 2. Response surface diagrams for water-added ham products.
(a) Rigidity, (b) Elasticity, (c) Compromise metric.

In a yellow layer cake or madeira cake, the major functional
contributions to the product from the whole egg could be considered
to be foam stability, emulsification and gelation (26). Whey protein
concentrates, however, are also able to demonstrate these functional
properties (27). There are a number of broad similarities between
egg proteins and whey protein concentrates. The greatest
similarities between these two groups of proteins are in their
globular structure, their ability to undergo heat denaturation and
their significant content of sulphur containing amino acids.

A series of yellow layer cakes were prepared to evaluate the
performance of whey protein concentrate as an egg replacer by itself
and in combination with other interactive ingredients. The
formulation used for the control cake formulation containing whole
egg is shown in Table VIII. Whey protein concentrates were evaluated
by replacing the dried whole-egg power with WPC on a protein basis.

Cakes were assessed on the basis of crust characteristics, specific
volume and compressibility. Measurement of crust colour was carried
out using a Hunterlab Colour Quest. The `a' scale (red-green) was
found to be most appropriate for measuring the extent of crust
browning. Subjective assessments were made on other characteristics
of the cake crust. The specific volume was measured by rape seed
displacement and the compressibility by a penetration test.

Table IX shows the performance of these cakes in terms of their
product properties.

Table VIII.   Formulation of Yellow Layer Cake
Containing Whole Egg

Ingredient	Percentage
Butter shortening	13.0
Castor sugar	28.0
Salt	0.2
High ratio cake flour	25.8
Baking powder	1.2
Pregelatinized starch	0.5
Dried whole-egg powder	3.6
Colouring	0.125
Flavouring	0.175
Emulsifier	1.0
Water	26.4

Table IX. Performance of Cakes Using Egg and
Whey Protein Concentrate

Property	Protein Type	
	Whole Egg	WPC
Crust-colour	8	18
—characteristics	SM, NS, M	P, S, SH
Specific volume (cm/g)	3.36	3.18
Compressibility (mm)	4.7	3.8

SM = smooth	M = matt	S = sticky
NS - non-sticky	P = pitted	SH = shiney

The WPC-containing cake in comparison to the whole egg containing cake was characterized by a red-brown sticky crust, a slightly lower specific volume and a lower compressibility. The WPC-containing cake was not acceptable on the basis of the crust characteristics and the texture (compressibility).

Cakes prepared with a range of additional interactive ingredients and WPC were able to highlight some of the areas where the WPC was functionally deficient. After careful selection of ingredients a series of optimization trials were carried out and the performance of the best WPC-containing cake is shown in Table X.

Table X. Performance of Cakes Using Additional
Ingredients and WPC

Property	Whole Egg	WPC	Additional ingredients and WPC
Crust-colour	8	18	10.5
—characteristics	SM, NS, M	P, S, SH	SM, NS, M
Specific volume (cm/g)	3.36	3.18	2.79
Compressibility (mm)	4.7	3.8	5.6

Here it is apparent that the cake formulation with the added ingredients has brought about a substantial improvement to the crust characteristics. The specific volume though, has been reduced to an unacceptable level with significant collapse on cooling and the texture of the cake has become crumbly as indicated by the compressibility data.

A further series of ingredients were investigated for their ability to interact in the cake system and overcome the low specific volume and the high compressibility.

The performance of these cakes is shown in Table XI.

Table XI. Performance of Cakes with Best WPC-Containing
Formulation

Property	Whole Egg Formulation	Best WPC-Containing Formulation
Crust-colour	8	10.5
-characteristics	SM, NS, M	SM, NS, M
Specific volume (cm/g)	3.36	2.96
Compressibility (mm)	4.7	4.8

The cake resulting from the best WPC-containing formulation is acceptable albeit that the specific volume is a little lower than optimum. These results show that it is possible to replace whole egg in a yellow layer cake or madeira cake with a WPC-based ingredient, and produce an acceptable product. This has been achieved only through the optimization of interactions in the cake system. In this case an iterative procedure was required. Here the initial formulation optimization did not produce all the required product properties and the optimization process needed to be repeated to arrive at the acceptable formulation.

Conclusions

These two examples of the application of milk protein ingredients in food systems illustrate why these products are used and perhaps give an indication of the mechanisms which control their usage. In both cases the functionality exhibited by the dairy protein in the food system in question has been modified by other ingredients present in the food product.
It is accepted that the functional properties exhibited by proteins in simple systems do not always translate well to their use in formulated food systems. The most immediate challenge to the scientists and technologists working in this area is to narrow the gap between the fundamental science of functional property elucidation and the technology of protein application into food systems. From the technologists point of view it is important that the roles which ingredients play in food systems are more clearly understood. From the fundamental scientist's point of view it is important that functional property evaluations are extended beyond simple systems to include other typical food ingredients so that the extent of ingredient interactions can be quantified.
In conclusion then, the roles of milk proteins in food systems are many and varied. Their wide use within the food industry has stemmed from both the range of functional properties which they exhibit and also their continued success in promoting desirable properties in formulated food systems. Much still remains to be learnt though, in attempting to bring about the convergence of these two areas of milk protein research.

Acknowledgments

The original research work referred to in this paper was carried out within the Food Ingredients Section at the New Zealand Dairy Research Institute. The authors would like to thank Miss Sue Croft, Ms Melanie Gordon, Mr Rowland Cocup and Mr Tony Fayerman for supplying original material for publication.
The authors would also like to thank Dr Lance Broad for his contribution in carrying out statistical analyses.

Literature Cited

1. Sanderson, W. B.   Bulletin of the International Dairy Federation. 1988, No. 224 33-35.
2. Southward, C. R.; Walker, N. J.   In CRC Handbook of Processing and Utilization in Agriculture; Wolff, I. A. Ed.; CRC Press, Boca Raton, Fla. 1982; Vol. 1, p 445.
3. Evans, E. W.   In Advances in Food Research; Academic Press Inc., 1987; 31, Chapter 4.
4. Mulvihill, D. M.; Fox, P. F.   Bulletin of the International Dairy Federation; 1987; No. 209, 3-11.
5. Leman, J.; Kinsella, J. E.   Critical reviews in food science and nutrition; 1989; 28 (2) 115-138.
6. Marshall, K. R.; Harper, W. J.   Bulletin of the International Dairy Federation 1988, No. 233, p 21-32.
7. Modler, W. H.; Jones, J. D.   Food Technology 1987, 41, 10 p 114.
8. Kirkpatrick, K. J.; Fenwick, R. M.   Food Technology 1987, 41, 10 p 58.
9. Towler, C.   New Zealand Journal of Dairy Science and Technology 1976, 11, 24.
10. Roeper, J.   New Zealand Journal of Dairy Science and Technology 1977, 12, 182.
11. Connolly, P. B.   U.S. Patent 4 376 072, 1982.
12. Ottenhof, H. A. W. E. M.   U.S. Patent 4 519 945, 1985.
13. Kinsella, J. E.   Critical reviews in Food Science and Nutrition 1976 p 219.
14. Morr, C. V.   in Developments in Dairy Chemistry; Fox, P. Ed.; Applied Science Publishers, London, 1982; Chapter 12.
15. Schut J.   The Role of Milk Proteins Added to Foods; The Dutch Dairy Bureau: Cort van Lindenstraat 7, 2288 E V Rijswijk, The Netherlands, 1980.
16. Harper, W. J.; Peltonen, R.; Hayes, J.   Food Prod. Dev., 1980, 14, 10 p 52-56.
17. Anon.   Bulletin of the International Dairy Federation, 1989 No. 239.
18. Cocup, R. O.; Sanderson, W. B.   Food Technology, 1987 41, 10 p 86.
19. Harper, W. J.   Proc. APV Symposium for the Soft Drink Fruit Juice; Dairy and Food Industry, Auckland, New Zealand, p 40.
20. Lim, D. M.   Leatherhead Food R.A. Scientific and Technical Survey, No. 120, 1980.

21. van den Hoven, M. Food Technology, 1987, 41, 10 p 72.
22. McDermott, R. L. Food Technology, 1987, 41, 10 p 91.
23. Kjaergaard, Jensen, G.; Ipsen, R. H.; Ilsøe, C.   Food Technology, 1987, 41, 10 p 66.
24. Baldwin, R. W. In Egg Science and Technology; AVI Publishing Co. Inc., Connecticut, 1977, Chapter 16.
25. Powrie, W. D. In Egg Science and Technology; AVI Publishing Co. Inc., Connecticut, 1977, Chapter 6.
26. Shepherd, I. S.; Yoell R. W.   In Food Emulsions; Ed.; S. Friberg; Marcel Dekker Inc., New York, Basel, 1976, Chapter 5.
27. De Wit, J. N. Neth. Milk Dairy J. 1984, 38.

RECEIVED August 14, 1990

Chapter 6

Significance of Lysozyme in Heat-Induced Aggregation of Egg White Protein

Naotoshi Matsudomi

Department of Agricultural Chemistry, Yamaguchi University,
Yamaguchi 753, Japan

Heat coagulation is one of the important functional
properties of egg white. The heat-induced interactions among the heterogeneous proteins in egg white
were investigated. The factors contributing to heat-
induced aggregation between ovalbumin and lysozyme were
examined to elucidate the mechanism of aggregation. The
heat-induced aggregation was due to an electrostatic
attraction and SH-SS interchange between the heat-
denatured protein molecules. It was found that the addition of native lysozyme into heat-denatured ovalbumin
forms insoluble aggregates. These results indicated that
lysozyme interacts electrostatically with the monomeric
molecule of fully unfolded ovalbumin. Ovotransferrin is
the most thermolabile protein in egg white protein. The
heat-induced aggregation of ovotransferrin with lysozyme
was examined. The aggregation was increased remarkably
in the presence of lysozyme. The heat-induced aggregation
was mainly due to electrostatic and hydrophobic interactions. In addition, lysozyme affected the rheological
properties of ovalbumin gel. We discuss significance of
lysozyme in the heat-induced aggregation of egg white.

Egg white is extensively utilized as a functional food material in
food processing. Heat coagulability is one of important functional
properties of egg white; therefore, the heat denaturation of egg
white and its component proteins has been studied by many investigators (1-8). However, heat-induced interaction among the heterogeneous
proteins in egg white has been little studied. It has been reported
(9,10) that ovalbumin and lysozyme, the major proteins in egg white,
can interact electrostatically in nature. Cunningham and Lineweaver
(11) have studied the inactivation of lysozyme by ovalbumin during
heating, and reported that lysozyme was rapidly inactivated and that
an insoluble precipitate was formed by heat treatment. They suggested
that the sulfhydryl groups of ovalbumin could be implicated in the

0097–6156/91/0454–0073$06.00/0

inactivation of lysozyme; however, the mechanism for heat-induced aggregation between these proteins has not been elucidated. Matsuda et al. (12) have studied the heat-induced aggregation of ovomucoid with lysozyme, and presumed that ovomucoid and lysozyme molecules were brought close together by the electrostatic attractive force, unfolded by heating, and then aggregated through intermolecular forces. In this study, we examine the heat-induced interaction between lysozyme and ovalbumin or ovotransferrin and describe the significance of lysozyme in the heat-induced aggregation of egg white. This information can be used in preparing egg white protein blends with a range of gelling properties to supply the burgeoning food formulations sector of the food industry, and preparing non-thermal coagulating egg white protein to supply many culture mediums for microorganisms and animal tissue cells (13,14).

Materials and Methods

Preparation of Proteins. Ovalbumin was prepared from fresh egg white by a crystallization method in sodium sulfate (15). Lysozyme was prepared from fresh egg white by a direct crystallization method (16). Ovotransferrin was obtained from Sigma Chemical. The lysozyme-free egg white protein was prepared by treatment with Duolite C-464, cation exchange resin, according to the method of Li-Chan et al.(17).

Chemical Modification of Proteins. Succinylation of protein was carried out according to the procedures of Habeeb (18). Acetylation and citraconylation of lysozyme were carried out according to the method of Yamasaki et al. (19) and Nieto and Palacian (20), respectively. Reduction and carboxyamide methylation (RCAM) of ovalbumin was carried out as described by Crestifield et al. (21), using iodo-acetamide.

Measurement of Gel Hardness. Gels were made with 10% protein solution containing various concentrations of lysozyme by heat treatment for 30 min at 75°C or 80°C. The gel hardness was determined on the gel sections (6.0 mm diameter x 5.0 mm height) with an Instron Universal Testing Instrument (Model 112, Instron Co.), according to the method of Mulvihill and Kinsella (22). The gel section was compressed at 40% of its original height (2.0 cm). The gel hardness (gram) was calculated from the height of the force peak on the first compression cycle.

Measurement of Turbidity. A turbidometric method was used to study the interaction between lysozyme and ovalbumin or ovotransferrin. The turbidity development is used as a measure of the extent of interaction. In a typical experiment, 1.0 ml aliquots of ovalbumin solution (0.2% in 35 mM potassium phosphate buffer at pH 7.6) were treated with increasing amounts of lysozyme (0.2% in the phosphate buffer) and the total volume was made up to 2.0 ml by adding the buffer. In the case of ovotransferrin-lysozyme interaction, 0.1% protein solutions were used. Heat treatment was carried out by heating at the rate of 2°C per minute from 20° to 90°C, and the turbidity was measured at 540 nm. The heat denaturation of ovalbumin was carried out with 2.0 ml portions of the protein solution (0.1%).

<u>Gel Electrophoresis</u>. Polyacrylamide gel electrophoresis in sodium dodecylsulfate (SDS) was performed in a 15% gel according to the method of Laemmli (23). The insoluble aggregate formed was collected by centrifugation. The precipitate was washed twice with the phosphate buffer and then dissolved in a given volume of a 10 mM Tris-HCl buffer (pH 6.8) containing 1% SDS, 25% glycerol, 1% 2-mercaptoethanol (2-ME) and 0.025% bromophenol blue. The protein bands were stained with 0.05% Coomassie brilliant blue and destained by diffusion in methanol-acetic acid-water (20:10:70, v/v). The concentration of proteins on the electrophoretic patterns was determined by measuring their color densities at 565 nm on a densitometer. The protein contents of ovalbumin and lysozyme were estimated from the standard curves of these proteins measured by the densitometer.

<u>Gel Filtration</u>. Gel filtration of heat-denatured ovalbumin was carried out by high-performance liquid chromatography on a TSK Gel G3000SW column (Tosoh, Manufacturing Co., Tokyo, 0.75 x 30 cm). A 20 ul portion of 0.1% protein solution was loaded on the column at a flow rate of 0.8 ml/min, using 0.2 M potassium phosphate buffer (pH 6.9) as an eluent. A UV monitor was used to monitor the effluent at 280 nm. The chromatogram was depicted by using a Shimadzu C-R3A Chromatopac with a chart speed of 5 mm/min.

Results and Discussion

<u>Heat Stability of Lysozyme-free Egg White Protein</u>. Lysozyme is a highly basic protein which accounts for 3.5% of the protein in egg white (24). It is known to form electrostatic complexes with other proteins in egg white, such as ovomucin (25, 26), ovomucoid (12), ovotransferrin (27) and ovalbumin (28, 29). Thus, lysozyme is thought to play important roles in the heat aggregation or coagulation of egg white. To elucidate the role of lysozyme, lysozyme-free egg white protein was prepared by treatment with Duolite C-464, cation exchange resin, according to the method of Li-Chan et al.(17). Figure 1 shows SDS-PAGE patterns of Duolite-treated egg white protein. From the result, lysozyme was confirmed to be reduced specifically from egg white.

Figure 2A shows the effect of heating temperature on the aggregation of Duolite-treated egg white (lysozyme-free egg white). The protein solutions (0.2%) in phosphate buffer (pH 7.6, ionic strength 0.1) were heated, and turbidity was measured at 540 nm. The turbidity of untreated egg white protein increased progressively with increasing heating temperature, while the lysozyme-free egg white protein did not show significant changes of turbidity, even at 85 °C. This indicates that the lysozyme-free egg white protein is much more stable to heat. To examine a significance of lysozyme in egg white, lysozyme was added to the lysozyme-free egg white protein, and then heated to 70 °C. Figure 2B shows the effect of added lysozyme. As lysozyme concentration increased, the turbidity greatly increased. One mg of egg white protein contains 34 ug of lysozyme (30). As shown by an arrow in Fig. 2B, the turbidity at this point corresponded to that of native egg white protein (see Fig. 2A) This result indicated that the existence of lysozyme in egg white is very important for heat coagulation of egg white.

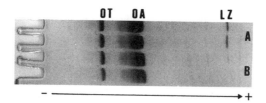

Figure 1. SDS-PAGE Patterns of Duolite-treated Egg white.
A, untreated egg white; B, Duolite-treated egg white;
OA, ovalbumin; OT, ovotransferrin; LZ, lysozyme.

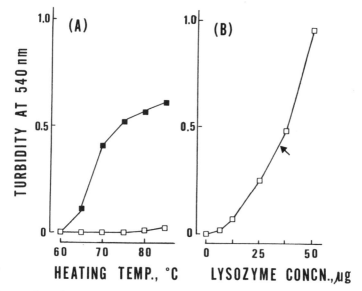

Figure 2. Effects of Heating Temperature and Added Lysozyme on
Aggregation of Duolite-treated Egg White (lysozyme-free egg
white). A, effect of heating temperature; B, effect of added
lysozyme; □, lysozyme-free egg white; ■, untreated egg white.

Heat-induced Aggregation between Ovalbumin and Lysozyme. When the dilute ovalbumin and lysozyme solutions (0.1%) were heated separately, the protein solution did not show any changes of turbidity, even at 90 °C. However, when mixed and heated, an insoluble aggregate was formed by heating to 70°C. Therefore, the heat-induced aggregate was examined by electrophoretic analysis. For the analysis of the aggregate, a mixture of ovalbumin and lysozyme (1:0.1 mg/mg) was dissolved in the phosphate buffer containing various NaCl concentrations and then heated to 75 °C. Figure 3 shows SDS-slab gel electrophoretic patterns of the precipitates. The precipitate obtained from the heated solution was found to consist of ovalbumin and lysozyme. The amount of precipitate was higher in the case of heating in the absence of NaCl, and decreased with an increase of ionic strength. The weight ratio of lysozyme to ovalbumin in the precipitates was assumed to be 1 in all cases, and the molar ratio of lysozyme (14,500) and ovalbumin (45,000) was estimated to be about 3.

The effects of ionic strength and heating temperature on the aggregation are shown in Fig. 4. The turbidity decreased with an increase of ionic strength, when compared at the same heating temperature. It has been reported by some investigators (10,11) that ovalbumin and lysozyme interacted more electrostatically in a salt-free solution at neutral pH, and that sodium and chloride ions shielded the charged groups in these proteins. The electrostatic interaction between ovalbumin and lysozyme is thus considered to be weakened by the addition of NaCl, and the formation of heat-induced aggregates would be depressed as the result of a decrease of electrostatic interaction. On the other hand, it has been reported (1,3) that the heat-induced aggregation of ovalbumin alone was affected by the net charge of protein molecules and inhibited by electrostatic repulsion among the protein molecules in the absence of salts. We have also reported (31) that, in the presence of NaCl, the soluble ovalbumin aggregates were readily formed during heat denaturation by hydrophobic interaction and disulfide bond formation, whereas the formation of the soluble aggregates was inhibited in the absence of NaCl. Therefore, in heterogeneous protein systems such as the mixture of ovalbumin and lysozyme, the heat-induced aggregation would be inhibited by a loss of the reactivity of ovalbumin to lysozyme due to the formation of soluble aggregates of ovalbumin itself as the concentration of NaCl was increased. The proteins did not show any changes in turbidity until heating to 60 °C, but became turbid when the heating temperature exceeded 70 °C. The heat-induced aggregation was observed at temperatures below the thermal transition point for heat denaturation of ovalbumin (76°C) and lysozyme (78°C) when determined from the changes in surface hydrophobicity and CD analysis (32). This indicates that the aggregation proceeded with the degree of heat denaturation of these proteins. It is suggested that the heat-induced aggregation may have been initiated by electrostatic interaction between the protein molecules and subsequently enhanced by hydrophobic interaction with heat denaturation. It is interesting that the extent of turbidity leveled off when samples were heated to 75 °C at lower ionic strength.

In order to examine the role of amino acid residues in the heat-induced aggregation, the effect of chemical modifications was studied. Figure 5 shows the effect of succinylated ovalbumin or lysozyme on the heat-induced aggregation. The heat-induced aggrega-

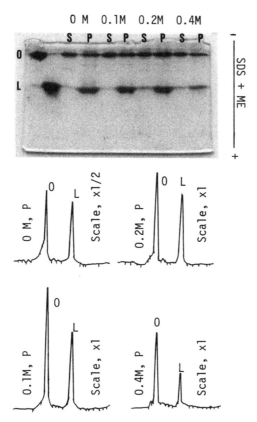

Figure 3. SDS Electrophoresis and Densitometric Scanning of
a Mixture of Ovalbumin (O) and Lysozyme (L) Heated in Various
Ionic Strengths at 75 °C. The numbers show NaCl concentration(M).
S, supernatant; P, precipitate.
(Reprinted with permission from ref. 9. Copyright 1986 Japan Society for Bioscience,
Biotechnology, and Agrochemistry.)

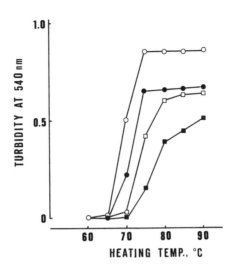

Figure 4. Effects of Ionic Strength and Heating Temperature on Heat-induced Aggregation between Ovalbumin and Lysozyme. The weight ratio of lysozyme to ovalbumin was 0.1. The concentration of NaCl as follows: O, no salt; ●, 0.1 M; □, 0.2 M; ■, 0.4 M. (Reprinted with permission from ref. 32. Copyright 1986 Japan Society for Bioscience, Biotechnology, and Agrochemistry.)

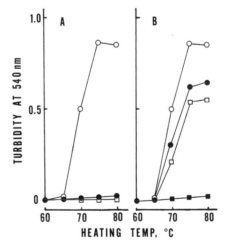

Figure 5. Effect of Succinylation of Amino Groups in Protein on Heat-induced Aggregation between Ovalbumin (OA) and Lysozyme(LZ). The weight ratio of lysozyme to ovalbumin was 0.1.
A: O, untreated OA - untreated LZ; ●, 85% modified OA - untreated LZ; □, untreated OA - 85% modified LZ.
B: O, untreated OA - untreated LZ; ●, 35% modified OA - untreated LZ; □, 52% modified OA - untreated LZ; ■, 85% modified OA - untreated LZ.
(Reprinted with permission from ref. 32. Copyright 1986 Japan Society for Bioscience, Biotechnology, and Agrochemistry.)

tion was depressed completely by the succinylation of ovalbumin or
lysozyme. Succinylation converted cationic amino groups to anionic
residues, and consequently the net charge of the succinylated
ovalbumin became negative. Therefore, it was presumed that
succinylated ovalbumin could interact electrostatically to a greater
extent with positively charged lysozyme at neutral pH. However, the
succinylated ovalbumin did not form aggregates with lysozyme during
heating. The heat-induced aggregation may therefore have been
depressed by the block of charged amino groups. The effect of the
degree of succinylation of ovalbumin on the heat-induced aggregation
is shown in Fig. 5B. As the extent of succinylation of ovalbumin
increased, the heat-induced aggregation with lysozyme reduced to a
greater extent. It is suggested from these results that the charged
amino groups (the ε-amino group) of ovalbumin and lysozyme played an
important role in the heat-induced aggregation between ovalbumin and
lysozyme. Figure 6 shows the effect of RCAM of ovalbumin on the
heat-induced aggregation. The mixture of RCAM-ovalbumin and lysozyme
was heated at 0.1 and 0.5 ionic strengths. The RCAM-ovalbumin significantly inhibited the heat-induced aggregation with lysozyme.
However, the heat-induced aggregates increased gradually with heating temperature. This result suggests that in addition to sulfhydryl
groups, electrostatic and hydrophobic interactions also participate
in the heat-induced aggregation. In addition, from the changes of
sulfhydryl groups in proteins during heating, it was suggested that
sulfhydryl-disulfide interchange reaction between ovalbumin and
lysozyme is involved in the heat-induced aggregation.

In conclusion, the scheme for heat-induced aggregation between
ovalbumin and lysozyme is illustrated in Fig. 7. When heated at a
low ionic strength, ovalbumin and lysozyme interact electrostatically at the molar ratio of 1 to 3 in the initial step of heating, and
in the next step, the two proteins are aggregated by hydrophobic
interaction and a sulfhydryl-disulfide interchange reaction. On the
other hand, when heated at a high ionic strength, ovalbumin itself
forms the soluble aggregates through hydrophobic interaction and
disulfide bonding in the initial stages of heating, and then
lysozyme is linked by a sulfhydryl-disulfide interchange reaction to
the soluble ovalbumin aggregates at the molar ratio of 3 to 1, consequently the complex becoming insoluble.

Aggregation between Lysozyme and Heat-denatured Ovalbumin. During
further experiments on the interaction between ovalbumin and
lysozyme, we have found that the addition of native lysozyme into
heat-denatured ovalbumin formed an insoluble aggregate, and that the
aggregation with lysozyme was depressed when ovalbumin was heated
above 75 C. In this section, the factors affecting the aggregation
were investigated.

The effect of the ratio of lysozyme to heat-denatured ovalbumin
on the aggregation was investigated by measuring the development of
turbidity. The ovalbumin solution (0.1%) was heated to 75 C, and
then lysozyme was added to the heat-denatured ovalbumin solution,
the results being shown in Fig. 8A. When the weight ratio of
lysozyme to ovalbumin was 0.02, the turbidity began to increase
slightly. The turbidity increased sigmoidally with an increase of
lysozyme concentration. Then, the resulting insoluble aggregates
were examined by electrophoretic analysis (Fig. 8B). The obtained

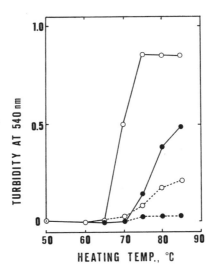

Figure 6. Effect of Carboxymethylation (RCAM) of SH Groups in Ovalbumin on Heat-induced Aggregation. The weight ratio of lysozyme to ovalbumin was 0.1. O, heated without NaCl; ●, heated with 0.4 M NaCl; ——, untreated ovalbumin; ---, RCAM-ovalbumin. (Reprinted with permission from ref. 32. Copyright 1986 Japan Society for Bioscience, Biotechnology, and Agrochemistry.)

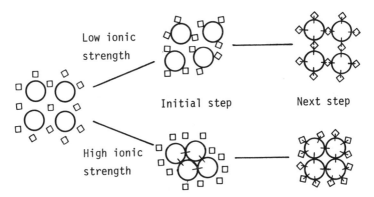

Figure 7. Scheme for Aggregation between Ovalbumin and Lysozyme during Heating. O, ovalbumin; □, lysozyme; ——, disulfide bond. (Reprinted with permission from ref. 32. Copyright 1986 Japan Society for Bioscience, Biotechnology, and Agrochemistry.)

precipitates were found to consist of ovalbumin and lysozyme. The molar ratio of lysozyme to ovalbumin in the precipitates was estimated to be about 1.5 in all cases, regardless of the concentration of adding lysozyme.

Figure 9 shows the effect of NaCl concentration on the aggregation. After the ovalbumin solution had been heated to 75 °C, salt and lysozyme were successively added to the heat-denatured ovalbumin solution. The turbidity greatly decreased as the concentration of the salt was increased. The formation of an insoluble aggregate was completely depressed by the addition of NaCl to a final concentration of 0.1 M. The interaction between the heat-denatured ovalbumin and lysozyme is therefore considered to be electrostatic, the formation of an insoluble aggregate being depressed as the result of a decrease of electrostatic interaction.

The effect of the degree of heat denaturation of the ovalbumin on the aggregation with lysozyme is shown in Fig. 10. The ovalbumin solutions were heated to various temperatures from 60 °C to 90 °C, and lysozyme was added to the heat-denatured ovalbumin solutions. The turbidity began to increase when ovalbumin was heated to 65 °C, and reached its maximum at 75 °C, before drastically decreasing after that. This phenomenon was also observed, when the mixture of ovalbumin and lysozyme was heated as described in the previous section (Fig. 4). Thus, the aggregation was found to be its maximum at the thermal transition temperature for the heat denaturation of ovalbumin, when determined from changes in surface hydrophobicity, CD analysis and exposure of sulfhydryl groups as previously described (32). We have proposed in the previous paper (31) that ovalbumin during heat denaturation was polymerized mainly by hydrophobic interaction and disulfide bond formation. The molecular properties of ovalbumin molecules during heat denaturation were therefore investigated in detail by HPLC analysis. Figure 11 shows the elution patterns of the supernatants obtained by centrifugation before and after the addition of lysozyme. Ovalbumin that was heated to 70 °C or 75 °C, as well as the unheated ovalbumin, gave a single peak with a retention time of 14.5 min. On the other hand, ovalbumin heated to 80 °C or 85 °C gave three peaks. These peaks seem to correspond to the monomer, dimer and trimer of the ovalbumin molecule. The fraction of the oligomer in the ovalbumin increased with the rise in heating temperature. As already mentioned for Fig. 10, the interaction with lysozyme decreased when ovalbumin was heated above 75 °C. Therefore, the interaction with lysozyme would be inhibited by polymerization of the ovalbumin molecule. It was suggested that lysozyme mainly interacted with the the monomeric molecule of fully unfolded ovalbumin by heating. The elution patterns of supernatants obtained by centrifugation after the addition of lysozyme to the heat-denatured ovalbumin are shown in Fig. 11B. Lysozyme gave a retention time of 20 min. The peak area of lysozyme in the supernatant was inversely proportional to the degree of turbidity. When ovalbumin was heated to 75 °C, the added lysozyme seemed to react almost entirely with the heat-denatured ovalbumin, because of no evidence of lysozyme in the supernatant. When ovalbumin was heated to 80 °C or 85 °C, a peak having a retention time of 8 min appeared in the supernatant by the addition of lysozyme. The peak area increased in proportion to the decrease in the oligomeric fraction of the ovalbumin molecule. This peak was found to consist of the

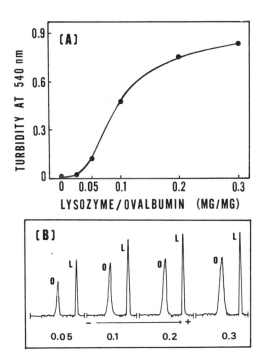

Figure 8. Effect of the Weight Ratio of Lysozyme to Ovalbumin on Aggregation and Densitometer Scans of SDS-Polyacrylamide Gels under 2-ME of Aggregates Obtained.   To 2 ml of a 0.1% ovalbumin solution (●) heated to 75 °C, increasing amounts of lysozyme were added.   The numbers in Fig. 8B show the weight ratio of lysozyme (L) to the heat-denatured ovalbumin.
(Reprinted with permission from ref. 33. Copyright 1987 Japan Society for Bioscience, Biotechnology, and Agrochemistry.)

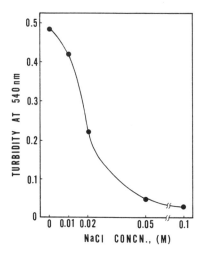

Figure 9. Effect of NaCl Concentration on the Aggregation between
Ovalbumin and Lysozyme. To 2 ml of a 0.1% ovalbumin solution
heated to 75°C, increasing amounts of NaCl were added, and then
lysozyme was added. The weight ratio of lysozyme to ovalbumin was
was 0.1.
(Reprinted with permission from ref. 33. Copyright 1987 Japan Society for Bioscience,
Biotechnology, and Agrochemistry.)

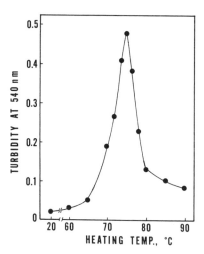

Figure 10. Effect of Heat-denatured Ovalbumin on the Aggregation
with Lysozyme. To 2 ml of a 0.1% ovalbumin solution heated to the
indicated temperature, lysozyme was added. The weight ratio of
lysozyme to ovalbumin was 0.1.
(Reprinted with permission from ref. 33. Copyright 1987 Japan Society for Bioscience,
Biotechnology, and Agrochemistry.)

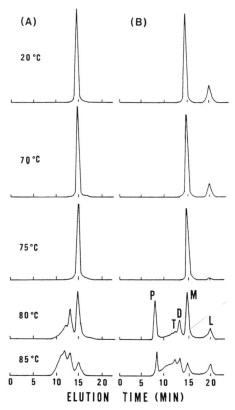

Figure 11. Elution Patterns of Supernatants Obtained by Centrifugation before and after Adding Lysozyme to the Heat-denatured Ovalbumin. (A) before adding lysozyme; (B) after adding lysozyme; P, ovalbumin polymer; T, ovalbumin trimer; D, ovalbumin dimer; M, ovalbumin monomer; L, lysozyme.

ovalbumin molecule alone. Therefore, the addition of lysozyme seems to have caused polymerization of the oligomeric ovalbumin molecules. This phenomenon has been known to be induced by the addition of NaCl. Therefore, the action of lysozyme is considered to have weakened the electrostatic repulsive forces between the denatured ovalbumin molecules and polymerized the oligomeric ovalbumin molecules by hydrophobic interaction and disulfide bond formation.

In conclusion, the aggregation between lysozyme and heat-denatured ovalbumin was caused by an electrostatic interaction at the molar ratio of 1.5. It was found that lysozyme interacted more with the monomeric molecule of the fully unfolded ovalbumin by heating, and formed insoluble aggregates with the ovalbumin. The added lysozyme also polymerized the oligomeric ovalbumin molecules to form the soluble ovalbumin aggregates. Thus, heat coagulation of egg white could be facilitated by interactions among such heterogeneous proteins as ovalbumin and lysozyme.

Heat-induced Aggregation between Ovotransferrin and Lysozyme. Among the constituent proteins of egg white, ovotransferrin is known to be one of the most heat-labile protein, and it is more unstable to heat at pH 7.5 than at pH 9.0 (34). The pH of egg white is also known to change from 7.3 to 9.7 during storage (35). Therefore, the heat-induced aggregation of egg white may be dependent on the denaturation of ovotransferrin.

Figure 12 shows the effect of heating temperature on the aggregation between ovotransferrin and lysozyme. The turbidity of ovotransferrin heated at pH 7.5 increased with an increase of heating temperature, and the formation of an insoluble aggregate was promoted by the addition of lysozyme (Fig. 12A). When heated at pH 9.0, ovotransferrin did not show any changes in turbidity, even at 85 C. However, when lysozyme was mixed and heated, an insoluble aggregate was formed by heating above 70 C (Fig. 12B). It is known that iron-binding ovotransferrin is much more stable to heat denaturation than the metal-free one (36). However, the heat-induced aggregation of iron-binding ovotransferrin was also promoted remarkably by adding lysozyme, at the both pHs.

Figure 13 shows the effect of NaCl concentration on the aggregation. In the case of pH 7.5, samples were heated to 60 C (Fig. 13A). The turbidity greatly decreased as the concentration of the salt was increased. The formation of insoluble aggregate of ovotransferrin itself occurred by heating to 60 C at pH 7.5, was also inhibited by the addition of NaCl. At pH 9.0, samples were heated to 75 C (Fig. 13B). The formation of an insoluble aggregate was completely depressed by the addition of NaCl to a final concentration of 0.5 M. In addition, the insoluble aggregates formed by adding lysozyme were dissociated completely into its components by SDS. Thus, the interaction between ovotransferrin and lysozyme was considered to be electrostatic and/or hydrophobic forces.

Effect of Lysozyme on Heat-induced Gelation of Ovalbumin. Figure 14 shows the effect of added lysozyme on hardness of ovalbumin gel. As lysozyme concentration increased, the gel hardness decreased gradually, and the transparency of gel became more turbid. Gel properties of protein are affected by the mode of denaturation and aggregation of protein. For the formation of the highly ordered gel

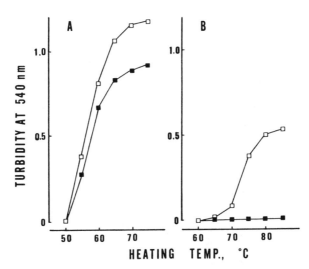

Figure 12. Effect of Heating Temperature on Aggregation between Ovotransferrin and Lysozyme. A, heated at pH 7.5; B, heated at pH 9.0; □, mixture of ovotransferrin and lysozyme; ■, ovotransferrin only.

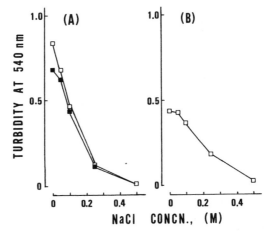

Figure 13. Effect of NaCl Concentration on the Aggregation. A, heated to 60°C at pH 7.5; B, heated to 75°C at pH 9.0; □, mixture of ovotransferrin and lysozyme; ■, ovotransferrin only.

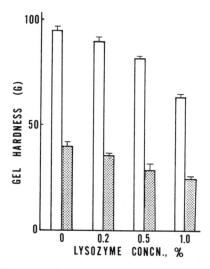

Figure 14. Effect of Lysozyme on Gel Hardness of Ovalbumin.
□, heated for 30 min at 80°C; ▣, heated for 30 min at 75°C.

matrix, it is imperative that the aggregation step proceeds at a slower rate than the unfolding step (37, 38). Thus, it was suggested that an ordered network formation of ovalbumin may be inhibited by the added lysozyme, because of the rapid formation of disordered aggregates of ovalbumin with lysozyme during heating.

Conclusions. In commercial food processing, the interaction of heterogeneous proteins is an important factor in determining the structure and texture of end products, and the capacity of proteins to aggregate or coagulate under practical conditions is a usual functional property in many systems. We described the significance of lysozyme in heat-induced aggregation of egg white protein. In the initial step of heating between lysozyme and ovalbumin, lysozyme interacts electrostatically with the monomeric molecules of fully unfolded ovalbumin, and in the next step, the two proteins are aggregated by hydrophobic interaction and a sulfhydryl-disulfide interchange reaction or lysozyme interacts with the soluble ovalbumin aggregates through intermolecular forces such as disulfide bonds and hydrophobic forces. Thus, heat aggregation of egg white could be facilitated by interactions among such heterogeneous proteins as lysozyme and ovalbumin or ovotansferrin. Such a interaction might affect greatly the gel properties of egg white.

Literature Cited

1. Holme, J. J. Phys. Chem. 1963, 67, 782.
2. Seideman, W. E.; Cotterill, O. J.; Funk, E. M. Poultry Sci. 1963, 43, 406.
3. Nakamura, R.; Sugiyama, H.; Sato,Y. Agric. Biol. Chem. 1978, 42, 819.
4. Hegg, P.-O.; Martens, H.; Lofgvist, B. J. Sci. Food Agric. 1979, 30, 981.
5. Shimada, K.; Matsushita, S. J. Agric. Food Chem. 1980, 28, 409.
6. Egelandsdal, B. J. Food Sci. 1980, 45, 570.
7. Ma, C.-Y.; Holme, J. J. Food Sci. 1982, 47, 1454.
8. Watanabe,K.; Matsuda,T.; Nakamura, R. J. Food Sci. 1985, 50,507.
9. Klotz, I. M.; Walker, F. M. Arch. Biochem. 1948, 18, 319.
10. Nakai, S.; Kason, C. M. Biochim. Biophys. Acta 1974, 351, 21.
11. Cunningham, F. E.; Lineweaver, H. Poultry Sci. 1967, 46, 1471.
12. Matsuda, T.; Watanabe, K.; Sato, Y. J. Food Sci. 1982, 47, 637.
13. Nishikawa, Y.; Kawai, F.; Mitsuda, H. J. Jap. Soc. Nutr. Food Sci. 1984, 37, 129.
14. Nishikawa, Y.; Kawai, F.; Mitsuda, H. J. Jap. Soc. Nutr. Food Sci. 1985, 38, 191.
15. Kekwich, R. A.; Cannan, R. K. Biochem. J. 1936, 30, 227.
16. Alderton, G.; Ferold, H. L. J. Biol. Chem. 1946, 164, 1.
17. Li-Chan, E.; Nakai,S.; Shim, J.; Bragg, D. B.; Lo, K. V. J. Food Sci. 1986, 51, 1032.
18. Habeeb, A. F. S. A. Arch. Biochem. Biophys. 1967, 121, 652.
19. Yamasaki, N.; Hayashi, K.; Funatsu, M. Agric. Biol. Chem. 1968, 32, 55.
20. Nieto, M.A.; Palacian, E. Biochim. Biophys. Acta 1983, 749, 204.
21. Crestifield, A. M.; Moore, S.; Stein, W. H. J. Biol. Chem. 1963, 238, 622.
22. Mulvihill, D. M.; Kinsella, J. E. J. Food Sci. 1988, 53, 231.

23. Laemmli, U. K. Nature 1970, 227, 680.
24. Osuga, D. T.; Feeney, R. E. In Toxic Constituents of Animal Food Stuffs; Liener, I. E., Ed.; Academic: New York, 1974.
25. Cotteril, O. J.; Winter, A. R. Poultry Sci. 1955, 34, 679.
26. Garibaldi, J. A.; Donovan, J. W.; Davis, J. G.; Cimino, S. L. J. Food Sci. 1968, 33, 514.
27. Ehrenpreis,S.; Warner,R.C. Arch. Biochem. Biophys. 1956, 61, 38.
28. Forsythe, R. H.; Foster, J. F. J. Biol. Chem. 1950, 184, 377.
29. Nichol, L. W.; Winzor, D. T. J. Phys. Chem. 1964, 68, 2455.
30. Osuga, D. T.; Feeney, R. E. In Food Proteins; Whitaker, J. R.; Tannenbaum, S. R., Ed.; AVI: Connecticut, 1977; p 220.
31. Kato, A.; Nagase, Y.; Matsudomi, N.; Kobayashi, K. Agric. Biol. Chem. 1983, 47, 1829.
32. Matsudomi, N.; Yamamura, Y.; Kobayashi, K. Agric. Biol. Chem. 1986, 50, 1389.
33. Matsudomi, N.; Yamamura, Y.; Kobayashi, K. Agric. Biol. Chem. 1987, 51, 1811.
34. Matsuda, T.; Watanabe, K.; Sato, Y. J. Food Sci. 1981, 46, 1829.
35. Heath, J. L. Poultry Sci. 1977, 56, 822.
36. Azari, P.R.; Feeney, R. E. Arch. Biochem. Biophys. 1961, 92, 44.
37. Hermansson, A.-M. J. Texture Studies 1978, 9, 33.
38. Hermansson, A.-M. In Functionality and Protein Structure; Pour-El, A., Ed.; ACS Symposium Series No. 92; American Chemical Society: Washington, DC, 1979; p 81.

RECEIVED June 20, 1990

Chapter 7

Formation and Interaction of Plant Protein Micelles in Food Systems

M. A. H. Ismond, S. D. Arntfield, and E. D. Murray

Food Science Department, University of Manitoba, Winnipeg, Manitoba R3T 2N2, Canada

Certain protein molecules have the ability to self-associate into a thermodynamically stable micelle configuration. This micelle phenomenon is important in the isolation of protein in the native state. In addition, these protein micelles may represent an important functional intermediate in the formation of a 3-dimensional framework. Under certain conditions, protein micelles will interact to form networks of amorphorous protein sheets reminiscent of the protein structures in a variety of food products. This potential of proteins to dynamically self-associate has been examined using storage proteins from *Vicia faba* as models. In addition, the effect of various environmental parameters on micelle formation were examined with a view to establishing a critical hydrophillic-hydrophobic balance which appears to be necessary for micelle formation and interaction. That is, the attractive hydrophobic forces must be dominant over repulsive electrostatic interactions to allow the micelle phenomenon to occur.

Using a broad definition, a micelle is considered to be a particle resulting from the association of amphiphilic molecules containing both hydrophillic and hydrophobic moieties. The micellar aggregate results when these molecules are in an aqueous environment above a critical protein concentration (CPC) and are characterized as having an interior hydrophobic core while the hydrophillic residues are in contact with the bulk water. Similar aggregates, in which, the core results from the association of hydrophobic heads can be obtained when nonpolar solvents are used; these are referred to as inverted or reverse micelles.

From classical chemistry, such behavior is normally associated with surface active agents (or surfactants), particularly detergents like sodium dodecyl sulfate (SDS) and cetyltrimethylammonium bromide (CTAB). For these materials, the importance of micelle formation is related to their role in the solubilization process.

0097–6156/91/0454–0091$06.00/0

Protein Micelles

The fact that proteins are amphiphilic molecules makes them good prospects for micelle formation. The most well known protein micelles in the food area are the casein micelles. These colloidal particles are roughly spherical, but heterogeneous in nature and have been the subject of many investigations. Although these casein micelles may fit into the broad definition of micelles, they represent a somewhat unique association between proteins, in which minerals such as calcium and phosphorous also play a major role (1).

The ability of proteins to form micelles, however, is not limited to the casein proteins. In fact, many of the globular storage proteins in plants have the ability to form spherical micelle structures which have characteristics similar to those described by the classical definition of micelles. The phenomenon was first observed with proteins from fababean (*Vicia faba*) flours and concentrates, when a high salt (0.3 M NaCl) protein extract was diluted in tap water (2). The cloudy precipitate which formed was shown to contain a large number of discrete spherical protein structures similar to those in Figure 1A. Furthermore, when these micelles were left stationary under the appropriate conditions, they tended to coalesce and form a gelatinous mass, called a protein micellar mass (PMM), which represented a highly purified protein estimated to be farily native by the calorimetric analysis (2). In fact, the spontaneous formation of the micellar mass was the basis of a patented procedure for the isolation of proteins from plant sources (3,4). This is not the only example where micelles are used for protein extraction. The interaction between reverse surfactant micelles and proteins has been suggested as a technique for extracting native proteins from aqueous systems on a large scale basis (5). In this application, the interaction between the protein and the aqueous core of the reverse micelle, rather than the ability of the protein to form its own micelle, was critical to protein recovery.

Over the past decade, protein micelles have been observed from a number of protein sources including, field peas, oats, canola and wheat gluten; micelles from several of these sources are shown in Figure 1. The micelles can appear as a disperse population (Figures 1A, 1B and 1D) or grouped together on the periphery of a piece of particulate matter in the system as seen for the canola sample (Figure 1C). Interestingly, unlike micelles from other protein sources, the micelles observed with the wheat gluten (Figure 1D) were not prepared using the salt extraction and subsequent dilution step to reduce ionic strength as described in the patent (3,4). They appeared only briefly, after simply mixing the wheat flour with water. Unlike the casein micelles, protein micelles formed from these storage proteins appear to result from formation of a hydrophobic core leaving the majority of hydrophillic residues at the micelle surface (2). On this basis, the term "protein micelle" seemed most appropriate to describe these structures.

Interaction of Protein Micelles

Protein micelles are somewhat more complex than detergent micelles in that protein structure does not permit complete orientation of the hydrophobic and hydrophillic residues. One of the results of this is the fact that protein micelles can interact with each other to form extensive networks, such as that shown in Figures 1E and 1F. These networks can be either three-dimensional structures (Figure 1E) or, in some instances, sheet like structures (Figure 1F). In fact, it is this network formation that

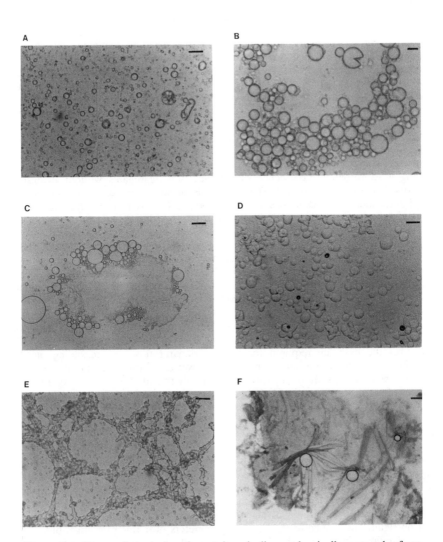

Figure 1.   Photomicrographs of protein micelles and micelle networks from various sources. A. fababean (bar represents 25μm) B. oat (bar represents 10μm) C. canola (bar represents 25μm) D. wheat gluten (bar represents 25μm) E. protein network from interactions of fababean micelles (bar represents 25μm) and F. protein sheet from interaction of field pea micelles (bar represents 15μm).

is critical to the formation of a protein isolate using the patented dilution procedure (3,4). However, not all conditions promote micelle formation and interaction. In fact, there is a great deal of variation in the size of micelles that can be produced as well as the extent of interaction between micelles. The extent of micelle interaction is key to optimizing the recovery of proteins using this isolation technique as well as producing a network which can be used directly in food systems. It is, therefore, necessary to understand the specific nature of the protein-protein interactions responsible for this phenomenon, so that the response can be predicted and manipulated to give the desired protein functional properties.

For this goal, the variation in micelle response has been rated using an arbitrary scale based on the visual appearance using light microscopy (6). Ratings ranged from 0 to 5, with 0 represent no micelle formation and 5 extensive network formation such as that seen in Figures 1E and 1F. Ratings 1 to 3 corresponded to micelles of increasing size from 2 to 20 μm; for example, Figure 1B corresponds to Rating 3. Rating 4 represented coalescence of micelles into aggregates without extensive network formation (not shown). The proteins from fababean were used to examine the interactions involved in this micelle response. In this respect, the two major storage proteins (legumin and vicilin) were isolated and subjected to a number of different environmental situations. Changes in the surface properties and conformation for the individual proteins were used to evaluate the influence of these different environments on the potential for micelle formation and interaction. Specifically, the surface hydrophobicity (So) of the protein was assessed using fluorescent probes, either cis-parinaric acid (CPA) (7) or 1-anilino-8-naphthalene sulfonate (ANS) (8). Conformation changes were reflected in the thermal properties assessed using differential scanning calorimetry (DSC). In this respect, the temperature associated with the thermal denaturation (Td) of the protein is an indicator of the stability of the molecule. The enthalpy of denaturation (ΔH) measures the enthalpy required for thermal denaturation; any unfolding of the protein due to the solvent should result in a lower ΔH during thermal analysis (9). The inclusion of these thermal parameters not only provided some indication of the impact of major conformational change on micelle formation but gave perspective to the observed changes in So in terms of the molecular associations involved.

Influence of pH on Protein Micelles

Variations in micelle response were examined at pH values between 5.5 and 9.0 for legumin and between 6.0 and 8.0 for vicilin. For both these proteins, an increase in pH beyond pH 6.0 resulted in a sharp decrease in the micelle response (MR) such that the highly interactive networks obtained at lower pH values were reduced to small non-interactive micelles at higher pH values. For legumin, Figure 2A, there was no significant change in the thermal parameters over this pH range, but there was a significant increase in the So value at pH 7.0 and above. It would appear that the higher pH values were sufficient to alter the balance of hydrophobic and hydrophillic residues on the surface of the protein and thus inhibit micelle interaction, without producing any major conformational changes. For vicilin, Figure 2B, the reduced micelle response was accompanied by a reduction in both So and Td values; the variations in the ΔH values in this pH range were not significant. In this case, it appeared that the slight destabilization of the protein structure resulted in increased burial of hydrophobic residues, possibly due to distortion of the subunit association at the quaternary level.

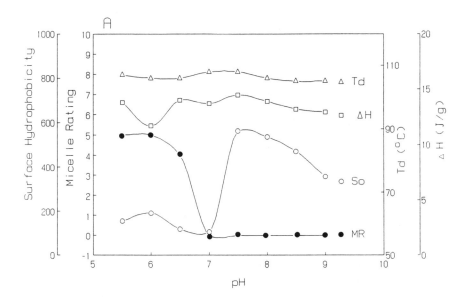

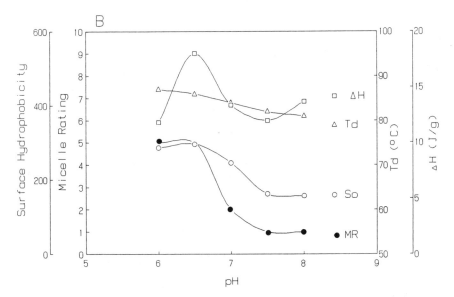

Figure 2. Changes in denaturation temperature (Td - △), enthalpy of denaturation (ΔH - □), surface hydrophobicity (So - o) in relation to micelle forming capacity (micelle rating - MR - ●) as a function of pH (0.1 M phosphate buffers). A. legumin (So determined with ANS) B. vicilin (So determined with CPA).

At the low ionic strength used in these studies (0.01M), the increase in pH away from the protein's isoelectric point increased the overall negativity of the molecule. This increase in surface charge may have been sufficient to exert a repulsion among established micelles and thus prevent further micelle association. In view of the fact that surface hydrophobicity increased for legumin but decreased for vicilin, the influence of hydrophobic interactions was minimal compared to the more obvious effect of charge repulsion. Although the association of hydrophobic residues may be key to both micelle formation and interaction, the high surface charge on the protein more than counteracted this attractive hydrophobic force.

Influence of Salts on Protein Micelles

To pursue further the relative importance of hydrophobic interactions in the coalescence of micelles, it was necessary to use environments which allowed manipulation of the surface hydrophobicity. Salts can provide such a tool in that protein conformation is affected by both salt type and concentration. There are two mechanisms which must be recognized when examining the influence of salt on proteins. At low concentrations (μ < 0.5), electrostatic influences, related to the polyionic nature of the protein are dominant (10). At higher concentrations, the position of salts (and usually the anions of the salts) in the Hofmeister or lyotropic series becomes important. The capacity of anions to exert a stabilizing influence at these concentrations has been related to their ability to increase the surface tension of the aqueous protein environment (11) as well as cause preferential hydration of the protein molecule (12). In this respect, a relationship has been established between the molal surface tension increment (σ) of a salt and its ability to stabilize proteins; the σ value can therefore be used to define the position of an anion (or cation) in the lyotropic series.

Several anions of sodium salts were chosen to investigate the influence of hydrophobic interactions on micelle formation and interaction; these salts included NaSCN (σ = 0.60 x 10^{-3} dyn g cm^{-1} mol^{-1}), $NaC_2H_3O_2$ (σ = 1.27 x 10^{-3} dyn g cm^{-1} mol^{-1}), NaCl (σ = 1.64 x 10^{-3} dyn g cm^{-1} mol^{-1}), $NaC_6H_5O_7$ (σ = 3.27 x 10^{-3} dyn g cm^{-1} mol^{-1}). A relationship between the thermal properties of the proteins and the position of the salt in the lyotropic series has been demonstrated at concentrations of 0.5 M and higher (13). At a concentration of 0.5 M, however, the influence of the σ values on protein stability (Td values) was minimal for legumin (Figure 3A). The NaSCN, which is known to destabilize proteins by binding of the thiocyanate anion, produced a significantly lower Td value while there was no difference in Td values for the other three salts in the series. Although there was considerable variation in the ΔH values, they could not be related to the σ of the salt. Nevertheless, as σ increased, there was a significant decrease in the surface hydrophobicity of the protein (So value) which implied there were minor conformational changes in the protein which resulted in increased burial of hydrophobic residues. This decrease in So was reflected in the ability of the micelles to interact. The lower So values associated with high σ values promoted the formation of distinct micelles (MR = 2) but did not promote micelle interaction as was seen with salts producing lower σ values.

For vicilin (Figure 3B), the stabilizing influence of the salts in the lyotropic series was evident by the increase in Td value with increasing molal surface tension increment. A slight increase was also observed for the ΔH values. This increased stability is probably associated with increased burial of hydrophobic residues, a

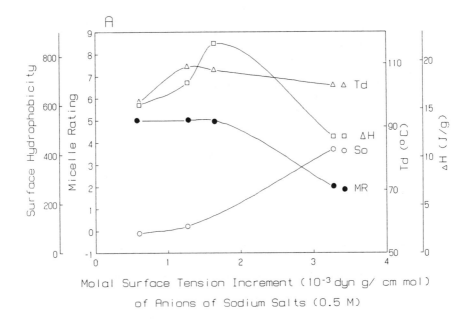

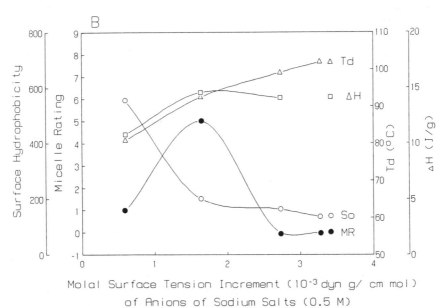

Figure 3. Changes in denaturation temperature (Td - △), enthalpy of denaturation (ΔH - □), surface hydrophobicity (So determined using ANS - ○) in relation to micelle forming capacity (micelle rating - MR - ●) as a function of the molal surface tension increment due to anions of sodium salts at a concentration of 0.5 M. A. legumin B. vicilin.

phenomenon which was reflected in the So values. In NaSCN, which causes protein destabilization, the So was very high (> 500). In the presence of stabilizing salts, however, So values were lower and a gradual decrease in So was observed with increasing σ values. Interestingly, the ability to form highly interactive micelle networks was associated only with intermediate σ values. The high So value associated with the NaSCN environment permitted micelle formation (MR = 2) but there was no evidence of further interaction. Unfolding of the molecule in this environment (low Td and ΔH values) may have changed the electrostatic profile sufficiently to provide charge repulsion between micelles. At the other end of the scale, the low So values associated with the chloride and citrate environments allowed only the formation of very small micelles (MR = 1). It would appear that the availability of hydrophobic residues is critical for the formation and coalescence of protein micelles in situations where the electrostatic profile is not prohibitive to further interaction.

Influence of Sucrose on Protein Micelles

Another way in which the surface hydrophobicity of a protein can be manipulated is through the use of sugars, such as sucrose. Protein stability is increased in the presence of sugars due to preferential hydration of the protein (14). The stabilizing influence of increasing sucrose content on vicilin was evident by the increase in Td and to a lesser extent ΔH values (Figure 4). Although the initial addition of sucrose caused a significant drop in the surface hydrophobicity, further increases in sucrose levels did not result in any change in So values. This would suggest that increased vicilin stability in the presence of sucrose was due to an increase in the cohesive force of the solvent rather than an increased burial of hydrophobic residues. The micelle response indicated that low levels of sucrose did not prohibit the formation of interactive micelle networks despite the decrease in the surface hydrophobicity. At higher sucrose concentrations, however, the micelles formed were unable to associate into extensive networks, even though there was no significant change in the surface hydrophobicity. It would appear that, in addition to the direct influence of hydrophobic and electrostatic interactions, changes in the cohesive force of the solvent can limit micelle coalescence. The energy required to promote cavity formation for the associated micelle structure may have been sufficient to prohibit the interaction process.

Influence of Urea on Protein Micelles

When working with storage proteins, denaturation conditions are often associated with protein aggregation. This results from electrostatic attractions between proteins. As hydrophobic interactions appear to play an important role in protein micelle formation, the influence of major changes in protein conformation on micelles is not so evident. The increased exposure of hydrophobic residues with protein unfolding is offset by a corresponding increase in the number of charged residues. To establish the role of protein conformation in the formation and interaction of protein micelles, increasing concentrations of urea were added to solutions of both legumin and vicilin (Figure 5).

The impact of urea on protein stability is evident; Td and ΔH values for legumin decreased at urea concentrations greater than 2.0 M, while significant changes in the thermal properties for vicilin were seen at urea concentrations of 1.0 M and greater. The ability of urea to denature proteins is well documented and

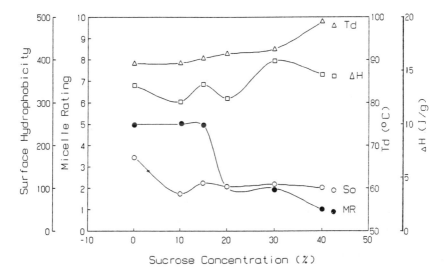

Figure 4.   Changes in denaturation temperature (Td - Δ), enthalpy of denaturation (ΔH - □), surface hydrophobicity (So determined using CPA - ○) in relation to micelle forming capacity (micelle rating - MR - ●) for vicilin as a function of sucrose concentration.

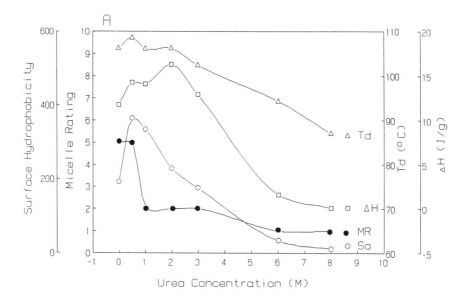

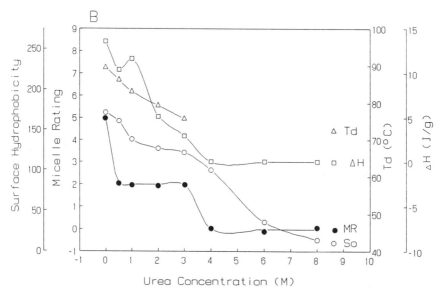

Figure 5.   Changes in denaturation temperature (Td - △), enthalpy of
denaturation (△H - □), surface hydrophobicity (So - o) in relation to micelle
forming capacity (micelle rating - MR - ●) as a function of urea concentration.
A. legumin (So determined with ANS) B. vicilin (So determined with CPA).

is related to an indirect decrease in hydrophobic interactions through increased solvation of nonpolar residues (15). This denaturation was evident at high urea concentrations where ΔH values approached zero and Td values were difficult or impossible to determine.

Changes in So values did not coincide with these major conformational changes. Even at low urea concentrations, where no major conformational changes were observed, the So values were altered. For legumin (Figure 5A), the addition of 0.5 M urea resulted in an increase in the surface hydrophobicity. It is conceivable that there was increased exposure of hydrophobic residues without major conformational change. At higher urea levels, however, the So decreased with increasing urea concentration, approaching a value of zero at 8.0 M urea. For vicilin (Figure 5B), even low levels of urea caused a decrease in the So value; higher urea concentrations again reduced the So values to levels approaching zero. The lower So values were unexpected as the deterioration of conformation to random coil would be paralleled by increased exposure of hydrophobic residues. The decreased surface hydrophobicity observed in these studies may reflect intermolecular aggregation as denaturation progressed. The effect of this loss in So on the micelle ratings was quite dramatic. For legumin, urea concentrations greater than 0.5 M resulted in a reduction in micelle interaction (MR = 2) and at a concentration which promoted extensive conformational change (6.0 M), the formation of micelles was further impaired. For vicilin, even low levels of urea inhibited micelle interactions and at levels of urea which promoted major conformational change, no micelle response was obtained.

Denaturation of storage proteins by urea can lead to aggregate formation which impairs the interactions of micelles. It would appear that electrostatically driven aggregation reduces the availability of hydrophobic residues necessary for micelle interaction and at high enough concentrations inhibits micelle formation. The decrease in both So values and MR, even at concentrations below those which promoted significant protein unfolding suggested that the influence of urea on the solubility of nonpolar residues may also limit the ability of the residues to form protein micelles, or interact with the hydrophobic probes used for the determination of surface hydrophobicity. Overall, it is clear that major conformational changes and increased solvation of hydrophobic residues are detrimental to micelle formation and interaction.

Noncovalent Forces in Protein Micelle Formation

It is evident that, like the native structure of the globular protein, protein micelles are dynamic structures which rely on a critical balance of attractive and repulsive forces. The ability of proteins to self-associate into a micelle arrangement is highly dependent on the surface properties of the protein, and as is the case with detergent micelles, is highly dependent on a distinct hydrophillic - hydrophobic balance.

As with the native protein, changes in the environment can inhibit the establishment of distinct protein micelles. Conditions which reduce the availability of surface hydrophobic residues (e.g. stabilizing salts) or decrease the ability of these residues to self associate due to increased reaction with the solvent (low urea concentrations) reduce the attractive forces so critical to the hydrophobic - hydrophillic balance. If, on the other hand, there is a change in the electrostatic profile of the protein either through pH manipulation or significant protein unfolding (high urea concentration), then the electrostatic forces may predominate. This could represent an increase in electrostatic repulsion within an established micelle population or the promotion of protein aggregation rather than micelle formation. In either case, the hydrophillic-hydrophobic balance has been disrupted and micelle network formation inhibited.

Micelle Networks in Food Systems

The potential for using the protein micelle systems has yet to be realized. The use of this micellization phenomenon as a technique for isolating protein is evident. In addition to giving a high yielding protein with minimal conformational change (2), many antinutritional factors are eliminated by this process (16). The formation of protein micelles from a wide range of plant storage proteins demonstrates the extent of this application.

The formation of a micelle network, however, should not be limited to its role in protein recovery. This type of interaction could be beneficial as a structural entity in food systems. In this respect, it is interesting to look at the wheat gluten system, where micelle formation was observed without a dilution step. It appeared as a natural product of gluten development. In view of the contribution of hydrophobic interactions to gluten functionality, it is possible that micelle networks may function in a similar, though not necessarily equivalent, capacity. In fact, in some preliminary investigations, a fababean micelle protein isolate has been used to replace gluten in bread with only a minimal loss in loaf volume. Further information on the macroscopic properties of these micelle networks is necessary to evaluate the extent to which they can be successfully incorporated into food products.

In making use of these networks, however, conditions must be carefully controlled to promote micelle formation and interaction. Conditions which promote major conformational change, enhance electrostatic interactions or repress hydrophobic association will diminish the integrity of these networks and cause a concomitant loss of protein functionality. With an appreciation for these factors, the future for micelle networks is most promising.

Literature Cited

1. Morr, C.V. In Functionality and Protein Structure; Pour-El, A. Ed.; ACS Symposium Series No. 92; American Chemical Society: Washington, D.C., 1979; pp 65-80.
2. Murray, E.D.; Myers, C.D.; Barker, L.D.; Maurice, T.J. In Utilization of Protein Resources; Stanley, D.W., Murray, E.D. and Lees, D.H., Eds.; Food and Nutrition Press, Westport, CT. 1981; pp 158-176.
3. Murray, E.D.; Myers, C.D.; Barker, L.D. Canadian Patent 1 028 522, 1978.
4. Murray, E.D.; Maurice, T.J.; Barker, L.D.; Myers, C.D. U.S. Patent 4 169 090, 1979.
5. Goklen, K.E.; Hatton, T.A. Biotech. Prog. 1985, 1, 69-74.
6. Ismond, M.A.H.; Murray, E.D.; Arntfield, S.D. Food Chem. 1986, 20, 305-318.
7. Kato, A.; Nakai, S. Biochim. Biophys. Acta 1980, 624, 13-20.
8. Hayakawa, S. and Nakai, S. J. Food Sci. 1985, 50, 486-491.
9. Arntfield, S.D.; Murray, E.D. Can. Inst. Food Sci. Technol. J. 1981, 14, 289-291.
10. von Hippel, P.H.; Schleich, T. In Structure and Stability of Biological Macromolecules; Timasheff, S.N.; Fasman, G.D., Eds.; Marcel Dekker Inc. New York. 1969; pp. 417-453.

11. Melander, W.; Horvath, C. Arch. Biochem. Biophys. 1977, 183, 200-215.
12. Arakawa, J.; Timasheff, S.N. Biochemistry 1982, 24, 6545-6552.
13. Ismond, M.A.H.; Murray, E.D.; Arntfield, S.D. Food Chem. 1986, 21, 27-46.
14. Ismond, M.A.H.; Murray, E.D.; Arntfield, S.D. Food Chem. 1988, 20, 189-198.
15. Creighton, T.E. Proteins, Structures and Molecular Properties; W.H. Freeman and Company: New York, 1984, p 149.
16. Arntfield, S.D.; Ismond, M.A.H.; Murray, E.D. Can. Inst. Food Sci. Technol. J. 1985, 18, 137-143.

RECEIVED May 16, 1990

Chapter 8

Diffusion and Energy Barrier Controlled Adsorption of Proteins at the Air—Water Interface

Srinivasan Damodaran and Kyung B. Song[1]

Department of Food Science, University of Wisconsin—Madison, Madison, WI 53706

To elucidate the influence of protein conformation on adsorption at liquid interfaces, the kinetics of absorption of ^{14}C-labelled ß-casein and structural intermediates of bovine serum albumin have been studied. Whereas the time course of adsorption of ß-casein followed square-root-of-time kinetics, which is consistent with a diffusion controlled adsorption phenomenon, the adsorption of BSA structural intermediates followed energy-barrier-controlled adsorption kinetics. These differences are interpreted in terms of the reversibility and/or irreversibility of adsorption, which in part is related to differences in molecular properties of proteins.

Adsorption of proteins at liquid interfaces and their subsequent behavior in the adsorbed state play a vital role in many biological and technological processes. The dynamics of protein adsorption at fluid interfaces is very different from that of a simple low-molecular-weight amphiphile. While in the case of small amphiphiles, such as aliphatic alcohols and alkyl sulfates, the entire molecule adsorbs and orients itself between the polar and nonpolar phases, the adsorption of protein macromolecules proceeds through sequential attachment of several polypeptide segments. In most cases, a greater portion of the molecule remains suspended in the aqueous phase and/or protruded into the non-aqueous phase in the form of 'loops' and 'tails' (1-4). In such a situation, the kinetics of adsorption and retention of the protein molecules in the adsorbed state during the initial stages of adsorption against thermal motions depends on the number of segments involved in the attachment and the sum total of the free energy of

[1]Current address: The Enzyme Institute, University of Wisconsin, Madison, WI 53706

adsorption of all the segments. It is conceivable that if this quantity is far greater than the thermal energy, the molecule would adsorb irreversibly to the interface.

The fact that adsorption of proteins proceeds through sequential attachment of several polypeptide segments implies that the kinetics and thermodynamics of adsorption of a protein are dependent on its conformational state in the solution phase. However, no systematic studies on the structure-adsorption relationship of proteins have been reported. To understand the role of protein conformation on adsorption, Graham and Phillips and co-workers (3,4) studied the kinetics and thermodynamics of adsorption of four structurally very different proteins, viz., lysozyme, bovine serum albumin, ß-casein and κ-casein at the air/water interface. The differences in the adsorption behaviors of these proteins were attributed to differences in their structural properties. However, it can be argued that the differences in the adsorption behaviors of these proteins may not solely be attributed to conformational differences alone, because the differences in their amino acid composition, distribution and charge characteristics would also influence their adsorption kinetics.

To elucidate the role of protein conformation on the kinetics of its adsorption at interfaces, recently we studied the adsorption behaviors of various structural intermediates of bovine serum albumin at the air/water interface (5). The rationale in that approach was that since neither the amino acid composition nor the sequence of the protein was altered, the differences in the adsorption behaviors of the intermediates could be attributed to their conformational differences alone. It was shown that the rate of adsorption of the serum albumin intermediates increased with the degree of unfolded state of the molecule in the bulk phase. Based on those observations, it has been suggested that the rate of adsorption of proteins at the air-water interface was not entirely diffusion controlled, but might also depend on the structure- dependent energy barrier to adsorption at the interface (5). In order to further understand the role of protein conformation on the adsorption process, in the present investigation we studied the kinetics of adsorption of [^{14}C]-labelled ß-casein and structural intermediates of serum albumin using the surface radioactivity method (6). The results show that the time course of adsorption of proteins is not a simple diffusion process, but a complex function of diffusion, subphase protein concentration and structure-dependent reversibility and irreversibility of adsorption at the interface.

Experimental

Materials. Crystallized and lyophilized bovine serum albumin (BSA) and ß-casein were purchased from Sigma Chemical Co. (St. Louis, MO). Na_2CNBH_3 and ultra pure (gold label) Na_2HPO_4 and NaH_2PO_4 were obtained from Aldrich Chemical Co. (Milwaukee, WI). [^{14}C]Formaldehyde was purchased from New England Nuclear Co. (Boston, MA). All other reagents used in this study were of reagent grade. Deionized and glass distilled water was used in all experiments.

Preparation of the Structural Intermediates of BSA. Structural intermediates of BSA were prepared as described previously (5,7). Two structural intermediates were prepared by blocking the sulfhydryl groups of the protein with iodoacetamide at 2 and 6 hr after initiation of the refolding process. The samples were dialyzed exhaustively against water at pH 7.0, lyophilized, re-dissolved in water at pH 7.0 and chromatographed on Sephadex G-25 to remove traces of bound iodoacetamide, and then finally lyophilized.

The BSA samples were [^{14}C] labelled by reductive methylation of the lysyl residues with H^{14}CHO according to the method of Matthews et al (8). To 20 mg of BSA dissolved in 20 ml of 20 mM phosphate buffer, pH 7.5, [^{14}C] formaldehyde was added (0.5 mM) followed by addition of 20 mM NaCNBH$_3$. The reaction mixture was stirred for 2 h at room temperature and then dialyzed exhaustively against water and lyophilized. About 2 to 3 lysyl residues in the protein were labelled under the above reaction conditions.

Circular Dichroism. The circular dichroic spectra of the BSA samples were determined using a modified and computerized Cary Model 60 spectropolarimeter (On-Line Instruments System, Inc., GA). A cell pathlength of 1 mm and a protein concentration of 0.01% were used. Each spectrum was an average of 10 scans. The secondary structure estimates were calculated from the CD spectra as described by Chang et al. (9).

Hydrodynamic Radius. The hydrodynamic radii of BSA and its intermediates were determined using a column chromatographic method (10). A Sephacryl S-300 column (100 x 1.6 cm) equilibrated with 20 mM phosphate buffer, pH 7.5 was used. A standard curve for elution volume versus hydrodynamic radius was obtained by eluting standard proteins whose hydrodynamic radii were known. These included ovalbumin (3.05 nm), thyroglobulin (8.5 nm), α-globulin (5.3 nm) and tyrosinase (4.23 nm) (10). The hydrodynamic radii of BSA and its structural intermediates were determined from their elution volume. The elution profile was monitored by measuring the absorbance at 280 nm.

Adsorption Studies. The rate of change of surface concentration of ^{14}C-labelled BSA at the air-water interface was monitored by the surface radioactivity method (6,11) using a rectangular gas flow counter with a mylar window (8 cm x 4 cm; Ludlum Instruments Inc., Sweetwater, TX) as the probe. The carrier gas was 98% argon and 2% propane. The surface pressure measurements were made using the Wilhelmy plate method using a Cahn electrobalance (Model 2000; Cahn Instruments Co., CA) (3,5,12).

In a typical experiment, the rectangular gas flow counter was setup at one end of a teflon trough (29 x 5.6 x 0.9 cm). A protein stock solution (about 0.1%) was prepared in 20 mM sodium phosphate buffer, pH 7.0. The temperature of the stock solution was brought to 30° C. An aliquot of this solution was diluted to the required final concentration with the same buffer preincubated at 30° C. The protein solution (140 ml) was gently poured into the teflon trough. Under these conditions the distance between the mylar window and the liquid surface in the trough was about 3 mm. The liquid surface was cleaned by gently sweeping with a fine capillary attached to an aspirator. The protein was then allowed to adsorb from the unstirred subphase to the air-water interface. The counts per minute (cpm) were integrated using a rate meter (Model 2200, Ludlum Instruments, Inc., Sweetwater, TX) for a specified time interval (usually 1 min during the initial period of adsorption, and 10 min at later stages) and printed out on a strip-chart calculator interfaced with the rate meter. To correct for the background radioactivity from the bulk phase, a standard curve relating cpm versus specific radioactivity was constructed using Na_2 $^{14}CO_3$ solutions. To convert the cpm into surface concentration in mg/m^2, a calibration curve relating cpm to surface concentration was constructed by spreading the [^{14}C]-labelled proteins on 1 M sodium sulfate solution. The surface concentration was increased by compressing the film. Preliminary experiments indicated that the reproducibility of the rate curves was within ± 0.1 mg/m^2.

Results and Discussion

BSA Intermediates. The kinetics of refolding of reduced and urea-denatured BSA and the method of preparation of the structural intermediates of BSA have been reported previously (5,7). Based on those studies, two structural intermediates of BSA were prepared by blocking the refolding process at 2 and 6 hr time intervals. The circular dichroic (CD) spectra of these two intermediates and that of the native BSA are shown in Figure 1 and some of the hydrodynamic and structural properties of these intermediates are given in Table I. The CD spectra of all the BSA samples exhibited ellipticity minima at 207 and 221 nm, which are indicative of α-helical structure. The secondary structure prediction from the CD spectra using the method of Chang et al (9) indicated that the intermediates 1 and 2 (entrapped at 2 and 6 hr time intervals, respectively) contained significant amount of β-sheet structure compared to that of the native BSA (Table I). The hydrodynamic radius of 3.68 nm for the native BSA obtained from gel filtration on sephacryl S-300 column was in good agreement with the published value of 3.55 nm (10). The unfolded BSA structural intermediates exhibited higher hydrodynamic radius; the hydrodynamic radius increased with the degree of unfolding and followed the order native < intermediate 2 < intermediate 1.

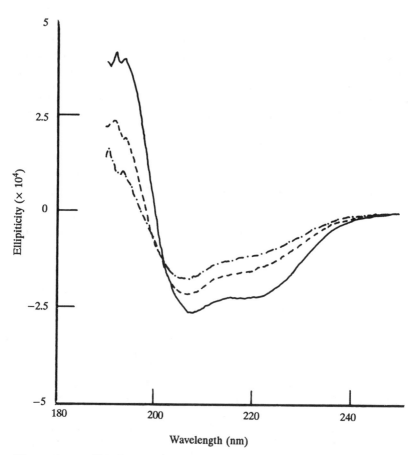

Figure 1:   Circular dichroic spectra of BSA intermediates in 20 mM
             phosphate buffer, pH, 7.0. Native BSA, ————;
             intermediate 1, —·—·—; intermediate 2, – – – –.

Table I. Conformational Characteristics of BSA Structural Intermediates

Protein	Hydrodynamic Radius nm	%α-helix	%β-sheet	%β-turns	% random
Intermediate 1	4.13	20	25	5	50
Intermediate 2	3.80	30	20	0	50
Native	3.69	55	0	0	45

Adsorption of BSA Intermediates at the Air/Water Interface. The rate of arrival of protein molecules at the interface from a dilute bulk phase is considered to be a diffusion controlled process and follows the relationship (13, 14),

$$d\Gamma/dt = C_o (D_s / 3.1416\, t)^{\frac{1}{2}} \qquad [1]$$

or, in the integrated form

$$\Gamma = 2 C_o (D_s / 3.1416)^{\frac{1}{2}} t^{\frac{1}{2}} \qquad [2]$$

where C_o is the bulk phase protein concentration, D_s is the diffusion coefficient, t is time and Γ is the surface concentration. A plot of Γ versus $t^{\frac{1}{2}}$ will be linear for a purely diffusion controlled process. The time courses of changes in the surface concentration during adsorption from dilute solutions of BSA intermediates are shown in Figure 2. The Γ versus $t^{\frac{1}{2}}$ plots were nonlinear for all the BSA intermediates at all bulk phase concentrations. This was more pronounced at higher concentrations. The rate of adsorption of intermediate 1 (which was trapped at 2 h of refolding) was much higher than that of either intermediate 2 or native BSA. The relative rates of adsorption apparently followed the order intermediate 1 > intermediate 2 > native BSA. These results apparently indicate that the kinetics of adsorption of BSA at the air-water interface was very much influenced by its solution conformation state.

The diffusion coefficients of native BSA and its intermediates, calculated from the linear region of the Γ versus $t^{\frac{1}{2}}$ plots at 1-10 min adsorption time interval, are given in Table II. For all the BSA samples the apparent diffusion coefficient decreased with increase of bulk phase concentration. Furthermore, the values of the apparent diffusion coefficients were about 2 to 3-fold higher than the conventional diffusion coefficient in bulk phase, which is about 0.6×10^{-10} m^2/sec (15). It should also be noted that at any given bulk phase

A

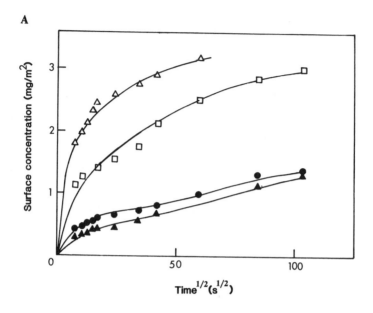

B

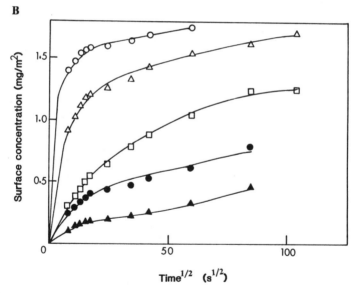

Figure 2:   Surface concentration versus $t^{\frac{1}{2}}$ plots of BSA
             intermediates. The protein concentrations were: ▲—▲,
             $5\times10^{-5}\%$, ●—●, $10^{-4}\%$; □—□, $2\times10^{-4}\%$; △—△, $5\times10^{-4}\%$,
             ○—○, $10^{-3}\%$ (not shown for intermediate 1). (A)
             Intermediate 1; (B) intermediate 2; (C) native BSA.

Continued on next page

C

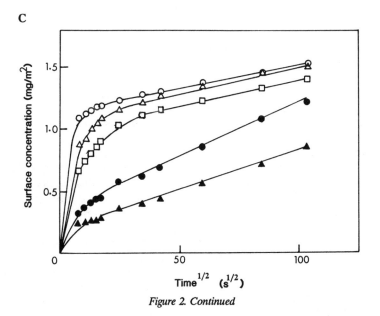

Figure 2. Continued

Table II. Diffusion Coefficients of BSA and ß-Casein

	$D \times 10^{10}$ m^2/sec			
C_o (wt %)	BSA Intermediate #1	BSA Intermediate #2	Native BSA	ß-Casein
5×10^{-5}	1.96	1.47	1.34	3.59
1×10^{-4}	1.48	1.07	1.07	2.98
2×10^{-4}	1.12	0.82	0.93	1.40

concentration, the apparent diffusion coefficient of intermediate 1 was greater than those of intermediate 2 and native BSA; this is unexpected because, since the hydrodynamic radius of intermediate 1 is larger than that of the other two BSA samples, one should expect a lower diffusivity for intermediate 1. The Γ and π obtained after an adsorption time of 15 h are plotted in Figure 3A and 3B, respectively, against the logarithm of bulk protein concentration C_o. Since true equilibrium was not reached at 15 h of adsorption, the curves in Figures 3A and 3B should be regarded only as apparent isotherms. In the case of intermediate 1, the surface concentration increased with the bulk phase protein concentration and reached a plateau in the range of 2×10^{-4} to 5×10^{-4}% (Figure 3A). This indicates that up to about 5×10^{-4}% bulk phase concentration, the intermediate 1 forms a monolayer at the air-water interface; above this concentration further adsorption seems to result in the formation of multilayers. In contrast to the behavior of intermediate 1, the shapes of the adsorption isotherms of intermediate 2 and native BSA indicate that, in the concentration range studied, these more folded structural forms of BSA form monolayers only at the air-water interface. At any given bulk phase concentration, the surface concentration of intermediate 1 was significantly higher than those of intermediate 2 and native BSA. For example, at $C_o = 5 \times 10^{-4}$% (the concentration at which all the three BSA samples form monolayers), the surface concentrations of intermediate 2 and native BSA were about 2.0 mg/m^2 and 1.5 mg/m^2, respectively, whereas that of intermediate 1 was about 3.5 mg/m^2. This apparently indicates that the state of conformation of the protein in the solution phase greatly affects the thermodynamics of its adsorption and conformation at the interface and thus the molecular surface area occupied (i.e., $1/\Gamma$) at the interface. In this regard, at the monolayer coverage, the relative surface area occupied by BSA apparently follows the order: intermediate 1 < intermediate 2 < native.

The surface pressure isotherms of the BSA intermediates are presented in Figure 3B. The shapes of the surface pressure isotherms were very similar for all the intermediates. However, at any given bulk phase

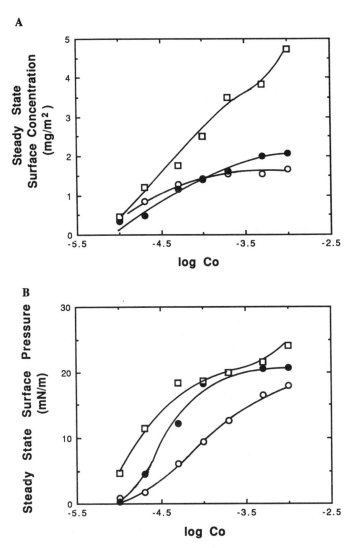

Figure 3:   Adsorption isotherm (A) and surface pressure isotherm
(B) of BSA intermediates. □—□, intermediate 1; ●—●,
intermediate 2; ○—○, native BSA.

concentration, the differences in surface pressure among the intermediates followed the order: intermediate 1 > intermediate 2 > native. Analysis of both the adsorption and surface pressure isotherms reveals the following information: Although the adsorption isotherms of intermediate 2 and native BSA were very similar (Figure 3A), the surface pressure isotherms were very different (Figure 3B). It is apparent that even though the surface concentrations of intermediate 2 and the native BSA were almost the same over the range $C_o = 1 \times 10^{-5}$ to 5×10^{-3}, the surface pressure of intermediate 2 was consistently higher than that of native BSA. This implies that even though the molecular areas $(1/\Gamma)$ occupied were almost the same, the conformation and the molecular orientation of the hydrophobic and hydrophilic residues at the interface were probably quite different for these two proteins. These differences might be responsible for the apparent differences in their influence on the surface forces. In the case of intermediate 1, the higher surface pressure at any given bulk-phase concentration might be due to higher surface concentration.

The data in Figures 3A and 3B suggest that in terms of surface activity per molecule, the conformational state of intermediate 2 (i.e., the intermediate entrapped at 6 hr during refolding) seems to exhibit higher surface activity than the other two structural states. This analysis is in good agreement with the previous conclusions, which were based on a different experimental approach (5). Furthermore, it is generally assumed that a highly unfolded protein molecule would occupy a greater area at the interface (3,4). However, the data presented in Figure 3A apparently do not agree with this viewpoint, because intermediate 1, which is highly unfolded, actually occupied lesser surface area than the other two BSA samples. The data, in fact, suggest that for a protein to occupy a greater surface area at the interface and exert a greater affect on the surface forces, it should possess folded three-dimensional structure.

Adsorption of ß-casein at the Air-Water Interface. To determine whether the nonlinearity of the Γ versus $t^{\frac{1}{2}}$ plots for BSA and its structural intermediates is in some way related to their molecular flexibility or inflexibility, the kinetics of adsorption of ^{14}C-labelled ß-casein, which is a flexible random coil protein, was studied. The Γ versus $t^{\frac{1}{2}}$ curves for the adsorption of ß-casein at three different bulk phase concentrations exhibited linearity up to about 2 h of adsorption (Figure 4). The apparent diffusion coefficients, calculated from the slopes of these straight lines, are given in Table II. The apparent diffusion coefficient decreased with increase of bulk phase concentration, which was very similar to the behavior observed for the BSA intermediates. Furthermore, the apparent diffusion coefficients were 2 to 5 times greater than that of the conventional diffusion coefficient, which is about 0.7×10^{-10} m²/sec.

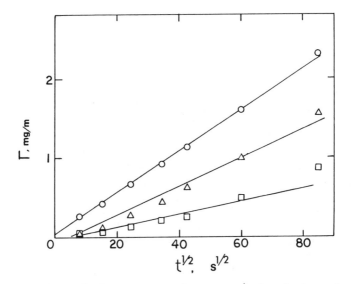

<u>Figure 4</u>:   Surface concentration versus $t^{\frac{1}{2}}$ plots for ß-casein adsorption at the air-water interface. □—□, $5\times10^{-5}\%$, △—△, 1×10^{-4}; ○—○, $2\times10^{-4}\%$.

Discussion

The mass transport of proteins from the bulk phase to an interface is considered to be a diffusion controlled process, and follows the relationship described by the Eqs. [1] and [2]. According to Eq. [2] the Γ-$t^{1/2}$ should be linear with $\Gamma = 0$ at $t = 0$. However, while these conditions are obeyed by ß-casein, they are not completely obeyed by BSA and its intermediates as shown in Figure 2.

Several investigators have suggested that the nonlinearity of the Γ-$t^{1/2}$ curves observed for various proteins might be related to progressive development of an energy barrier at the interface (3,13,16-18). That is, after a finite time of adsorption, the rate of arrival of the protein at the interface is

$$d\Gamma/dt = C_o \, k \, \exp(-E_a/kT) \qquad\qquad [3]$$

The energy barrier E_a is considered to be related to the surface pressure barrier $\pi \Delta A$ at the interface (16,17), where ΔA is the area that has to be cleared at the interface for the adsorption of protein molecules. MacRitchie (14) had pointed out that Eq. [1] is only valid for the very early stages of the adsorption process, i.e., when $\pi < 0.1$ mN/m. Above 0.1 mN/m surface pressure, because of the development of surface energy barrier, not every collision of the protein molecules leads to adsorption (14,17); this results in the nonlinearity of the Γ versus $t^{1/2}$ plots. This means that Eqs. [1] and [2] are not applicable to real situations in which the attainment of $\pi = 0.1$ mN/m takes place within seconds. However, the adsorption behavior of ß-casein, which shows a linear relationship between Γ and $t^{1/2}$ up to about 2 h, i.e., $\pi > 15$ mN/m (3), seems to disagree with this argument. Despite appreciable buildup of surface pressure, the adsorption of ß-casein follows a $t^{1/2}$ kinetics, whereas the BSA intermediates follow an energy barrier controlled kinetics. This significant difference might be due to two possible reasons: First, the area ΔA required for adsorption of ß-casein may be very small compared to that of BSA, so that the value of the exponential term in Eq. [3] is almost equal to unity even at moderately high surface pressure. A simple calculation indicates that at $\pi = 10$ mN/m, the value of ΔA should be less than 1 $Å^2$ in order for the exponential term to be unity. This is unreasonable, because one would expect that at least one amino acid residue ($\Delta A = 15 \, Å^2$) would be needed to anchor a protein at the interface. Even in such a case, the protein molecule should have high flexibility and should be able to undergo rapid conformational change/reorientation as it approaches the surface. DeFeijiter and Benjamins (19) pointed out that, because of steric constraints, it would be physically impossible for globular proteins to occupy only a small area at the interface. These analyses suggest that the differences in the adsorption kinetics of BSA and ß-casein cannot be explained or attributed to existence or nonexistence of an energy barrier arising from surface pressure.

Another probable reason for the differences in the kinetics of adsorption of BSA and ß-casein may be related to the reversibility or irreversibility of adsorption. The model described by Eq. [1] assumes that every collision of the protein molecule with the interface leads to adsorption. However, the success of every collision leading to adsorption and retention of the molecule at the interface should be related to the hydrophobicity/hydrophilicity of the protein surface. Proteins, such as ß-casein, that are highly flexible and hydrophobic will have a high probability of adsorption at the interface, whereas proteins that are highly rigid and hydrophilic (such as BSA) would have low probability of adsorption upon collision, and also might easily desorb from the interface. In the former case, the time course of adsorption would tend to follow a $t^{\frac{1}{2}}$ kinetics, whereas in the latter case, which is a reversible adsorption, the kinetics would follow an energy barrier controlled adsorption, that is (3)

$$d\Gamma/dt = k_1 C_o \exp(-E_a/kT) - k_2 \Gamma (-E_d/kT) \qquad [4]$$

where k_1 and k_2 are the adsorption and desorption rate constants, respectively, and E_a and E_d are the energy barriers for adsorption and desorption, respectively. In other words, the significant difference in the nature of the kinetics of adsorption of ß-casein and BSA might be attributed to irreversibility and reversibility, respectively. For reversible adsorption, Eq. [1] and [3] may not be applicable because the back diffusion created by desorption would destroy the concentration gradient between the surface and the sub-surface.

In general terms, the phenomenological rate law for the adsorption of proteins at an interface can be written as

$$d\Gamma/dt = k C_o t^{(n-1)} \qquad [4a]$$

which upon integration gives

$$\Gamma = (k C_o/n) t^n \qquad [5]$$

where k is the macroscopic rate constant, C_o is the bulk phase concentration and n is a constant. It should be pointed out that when n=0.5, Eq. [5] becomes Eq. [2]. Taking logarithm, Eq. [5] becomes,

$$\log \Gamma = \log (k C_o/n) + n \log t \qquad [6]$$

A plot of log Γ versus log t will be a straight line with a slope of n. From the intercept the rate constant k can be calculated.

Analysis of the data in Figure 2 according to Eq. [6] is shown in Figure 5. Unlike the Γ vs. $t^{\frac{1}{2}}$ plots, the log Γ vs. log t plots were linear up to 15 h. Several interesting features are apparent. It should be noted that the

A

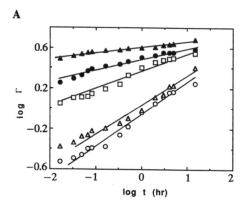

B

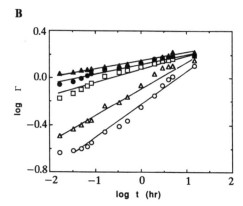

C

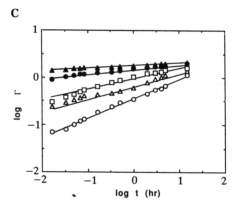

Figure 5:   The ln(Γ) versus ln(t) plots of BSA intermediates. The
protein concentrations were: ▲——▲, 5x10⁻⁵%; ●——●,
10⁻⁴%; □——□, 2x10⁻⁴%; △——△, 5x10⁻⁴%, ○——○, 10⁻³%.
(A) intermediate 1; (B) intermediate 2; (C) native BSA.

exponent of time n in Eq. [5] is not a constant, but is dependent on the bulk protein concentration; n increases with decrease of C_0 and tends to approach a value of 0.5 at very low C_0. A plot of the exponent n versus log C_0 for all the BSA intermediates approaches an asymptotic value of n = 0.5 at $C_0 < 10^{-6}\%$ (Figure 6). In other words, at $C_0 < 10^{-6}$ wt%, the back diffusion from the interface might be negligible, so that the maintenance of the concentration gradient between the surface and the subsurface apparently results in a diffusion controlled adsorption. At $C_0 > 10^{-6}$ wt %, the back diffusion from the interface may be very significant. The extent of deviation of n from the theoretical value of 0.5 can be regarded as a relative index of the reversibility of adsorption of the protein under a given set of experimental conditions.

The log Γ vs. log t plots at various C_0 appear to converge towards a constant surface coverage after about 100 h. This indicates that the protein adsorption isotherms based on 15 h or 24 h adsorption should be regarded only as apparent isotherms. It is also evident that in the range of C_0 studied, each of the BSA intermediates attain a constant surface coverage at infinite adsorption time, rgardless of the initial C_0; the extent of coverage, however, is dependent upon the conformation of the intermediate.

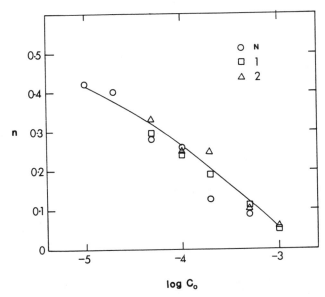

Figure 6:   Log C_0 versus n plots of BSA intermediates. The values of n were obtained from the slopes of the curves in Figure 5.

Conclusion

The mechanism of adsorption, retention and film formation of proteins at interfaces, such as air-water and oil-water interfaces, is very complex. Since food proteins act as surfactants in foam and emulsion type products, basic understanding of the molecular bases for protein adsorption at interfaces may provide novel strategies to improve the emulsifying and foaming properties of underutilized food proteins, such as soya proteins.

Acknowledgments

Support in part by the National Science Foundation Grant No. BCS-8913053 is gratefully acknowledged.

Literature cited

1. Benjamins, J.; DeFeijter, J.A.; Evans, M.T.A.; Grahams, D.E.; Phillips, M.C. Disc. Faraday Soc. 1975 59, 218-229.
2. Phillips, M.C.; Evans, M.T.A.; Graham, D.E.; Oladani, D. Colloid Polymer Sci. 1975 253, 424-427.
3. Graham, D.E.; Phillips, M.C. J. Colloid Interface Sci. 1979a 70, 403-408.
4. Graham, D.E.; Phillips, M.C. J. Colloid Interface Sci. 1979b 70, 409.
5. Damodaran, S.; Song, K.B. Biochim. Biophys. Acta 1988 954, 253-264.
6. Muramatsu, M. In Surface and Colloid Science, Matijevic, E., Ed.; vol. 6; Wiley: New York, 1973, p. 101.
7. Damodaran, S. Intl. J. Peptide Protein Res. 1989 27, 589-596.
8. Matthews, D.E.; Hessler, R.A.; Denslow, N.D.; Edwards, J.S.; O'Brien, T.W. J. Biol. Chem. 1982 257, 8788-8794.
9. Chang, C.T.; Wu, C.C.; Yang, J.T. Anal. Biochem. 1978 91, 13-31.
10. Tayyab, S.; Qasim, M.A. Biochim. Biophys. Acta 1987 913, 359-367.
11. Adams, D.J.; Evans., M.T.A.; Mitchell, J.R.; Phillips, M.C. Rees, P.M. J. Polymer Sci., Part C, 1971 34, 167.
12. Gaines, G.L., Jr., Insoluble Monolayers at Liquid-Gas Interfaces, Interscience: New York, 1966.
13. Ward, A.F.H.; Tordai, L. J. Chem. Phys. 1946 14, 453-461.
14. MacRitchie, F.; Alexander, A.E. J. Colloid Sci. 1963a 18, 453-457.
15. Tanford, C. Physical Chemistry of Macromolecules; Wiley: New York, 1961, p. 358.
16. MacRitchie, F.; Alexander, A.E. 1963b J. Colloid Sci. 1963b 18, 458-463.
17. MacRitchie, F. Adv. Protein Chem. 1978 32, 283-326.
18. Tornberg, E. J. Colloid Interface Sci. 1978 64,391.

19. DeFeijter, J.A.; Benjamins, J. In Food Emulsions and Foams; Dickinson, E., Ed.; Royal Society of Chemistry: London, 1987; p. 72.
20. Damodaran, S. and Song, K.B. In Surfactants in Solution; Mittal, K.L., Ed.; Plenum Press: New York, 1989; p. 391.
21. Perutz, M.F. Science 1978 201, 1187-1191.
22. Brown, J.R.; Shockley, P. In Lipid-Protein Interactions; Jost, P.; Griffith, O.H. Eds.;, vol.1; Wiley: New York, 1982; p. 25.
23. Pashley, R.M.; McGuiggan, P.M.; Ninham, B.W.; Evans, D.F. Science 1985 229, 1088-1089.
24. Israelachvili, J.N.; McGuiggan, P.M. Science 1988 241, 795-800.
25. Israelachvili, J.N.; McGuiggan, P.M. Nature (London) 1982 300, 341-342.

RECEIVED June 20, 1990

Chapter 9

Surface Activity of Bovine Whey Proteins at the Phospholipid–Water Interface

Donald G. Cornell

Eastern Regional Research Center, Agricultural Research Service,
U.S. Department of Agriculture, Philadelphia, PA 19118

The interaction of the major whey proteins
of cows' milk with phospholipids in a monolayer was studied with a Langmuir Film
Balance. The phospholipids were mixtures of
phosphatidylcholine and phosphatidyl-
glycerol. The ultraviolet and circular
dichroism spectra of transferred films were
used to determine the quantity and conformation of proteins interacting with the lipids
in the monolayer. The proteins were dissolved in the aqueous subphase and migrated
to the phospholipid/water interface where
adsorption was controlled by the solution pH
and calcium ion concentration. At neutral
pH, above the isoionic point of all proteins
studied, protein-phospholipid interaction
was minimal. Some binding of protein to the
lipids was observed when the subphase was at
the isoionic pH of the protein. Increased
lipid-protein binding was observed when the
subphase was below the isoionic pH of the
protein; under these conditions a film of
protein about one molecule thick formed
beneath the lipid monolayer. Calcium in the
subphase interfered with lipid-protein
binding, but the effect was pH dependent.
Circular dichroism spectra of transferred
films showed that the secondary structure of
the proteins was largely preserved in the
lipid-protein complexes. An electrostatic
mechanism for lipid-protein interaction is
discussed.

Dispersed fat particles in a stable system such as milk
are prevented from coalescing by repulsive forces between
their surface layers. Phospholipids assist in stabiliz-

ing the lipid phase of milk through formation of a bi-layer fat globule membrane (FGM) on the surface of the oil droplets (1). During milk processing steps such as homogenization, the fat globule particle size is deceased and the oil/water interfacial area is increased. The freshly exposed oil surface is able to adsorb additional surface active agents such as phospholipids and proteins to give a stabile emulsion. Since there is enough phospholipid in milk to form a film on a greatly expanded oil/water interface, this lipid undoubtedly plays an important role in stabilizing dairy and other food products which utilize homogenized milk. Although much work has been done on the surface activity of proteins at the oil/water interface, most of the efforts have centered on the interaction of proteins with neutral oil (hydrocarbons, triglycerides) surfaces; interactions with the polar lipids such as the phospholipids have received much less attention. Interaction between phospholipids and proteins depends to a degree on solution pH and mineral (principally calcium) content and on the properties of the lipid and proteins themselves. This of course implies that milk products, when used as a food additative, may exhibit different functional surface properties in different foods. Accordingly we decided to investigate the interaction of proteins with phospholipids over a range of pH's and calcium ion concentrations which were selected to include the conditions found in a variety of foods.

The caseins are the major milk proteins used to stabilize food emulsions, however whey is receiving increased attention as a food additive. This is especially true in acid foods where the whey proteins remain soluble. In such food uses, the end-use environment of whey can vary considerably ie. pH, salt, mineral content, etc. We are studying the interaction of the major whey proteins with film forming substances such as phospholipids under a range of solution conditions typical of food systems. In this work we report on the interaction of phospholipids in monolayers with the major whey proteins, beta-lactoglobulin (ß-LG), alpha-lactalbumin (α-LA), and bovine serum albumin (BSA). Mixtures of palmitoyloleoylphosphatidylcholine (POPC) and palmitoyloleoylphosphatidylglycerol (POPG) were chosen to simulate the zwitterionic (POPC) and charged (POPG) phospholipids of milk. The pH of the solutions ranged from 4.4 to 7; the calcium concentration was 0 to 8mM. These conditions include the ranges of pH and calcium ion (free, uncomplexed Ca^{2+}) found in many foods to which milk or dairy products might be added.

Experimental

Materials. The phospholipids were from Avanti Polar Lipids, Birmingham, AL; the proteins were from Sigma, St.

Louis, MO. The spreading solvent for the lipids was ACS Reagent Grade chloroform, treated to remove surface active impurities (2). Walpole's acetate (3) was used for pH 4.4 and 5.2, Tris-HCl for pH 7. A low buffer concentration (1-5 mM) was used to minimize light scattering in the films used for ultraviolet (UV) and circular dichroism (CD) spectroscopy.

The following are some relevant properties of the proteins used in this work. ß-LG; molecular weight (MW), 18,300; isoionic point (I_{IP}), 5.1; molar absorptivities (ϵ_M) for the backbone (190 nm) band, 1.56 X 10^6; mean residue absorptivity for the 190 nm band (ϵ_R), 9600 (7): α-LA; MW, 14,200; I_{IP}, 4.4; ϵ_M, 1.07 X 10^6; ϵ_R, 8700: BSA; MW, 66,300; I_{IP}, 5.2; ϵ_M, 5.06 X 10^6, ϵ_R, 8700. UV absorptivities for BSA and α-LA were determined in the authors' laboratory (unpublished observations).

A miniature Teflon trough (15 X 11 X 0.7 cm) with a dipping well (total depth 2 cm) was used for film transfer in this work. A magnetic stirrer was used to mix the protein in the subphase. The miniature trough was mounted in the cradle of a previously described (2) Langmuir balance from which the large trough had been removed. The barrier drive, Wilhelmy plate, constant-pressure control, and film transfer were all provided by the large film balance system. The experimental setup is illustrated schematically in Figure 1. A complete discussion of film balance techniques has been given (4).

Methods. The principle objective of this work was to estimate the amount and conformation of protein interacting with phospholipids at the lipid/water interface. The protocols for monolayer film and protein solution preparation, film and solution surface pressure measurements, and transfer of the monolayers for spectroscopic determinations have all been given in extenso (2, 5, 6); the following is a summary. The surface pressures (surface tension lowering) of the proteins were determined by the Wilhelmy plate technique on stirred solutions using thoroughly cleaned glassware. Techniques involving the formation and handling of lipid monolayers were carried out on the miniature film balance. A series of preliminary experiments were run to determine the optimum film pressure, protein concentration, etc., for UV and CD spectroscopy as follows:

(1) The surface pressures of ß-LG, α-LA, and BSA were determined as a function of concentration (up to about 80 mg/L) on buffered solutions. This established the maximum surface pressure that the proteins could achieve over the range of concentrations studied.

(2) Using the miniature film balance, monolayers of POPG/POPC (35/65 mol%) were formed by spreading the premixed lipid from chloroform solvent onto a

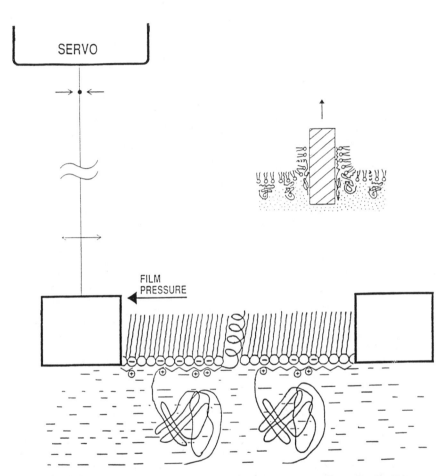

Figure 1.   Schematic diagram of a Langmuir-Blodgett
film balance showing proteins adsorbed from solution
onto the head groups of phospholipid monolayers.   The
inset shows a lipid-protein monolayer being transferred to the surface of a quartz plate for ultraviolet or circular dichroism spectroscopy. Eight
plates were used which gave a total of 16 monolayers
for spectroscopic investigations.

buffered subphase. After the spreading solvent
evaporated (15 min), the film was compressed to a
pressure greater than that achieved by the proteins
at the air/water interface in experiment (1) above.
Then, with the compression barriers stopped, protein
was injected into the subphase beneath the lipid
monolayer and the film pressure was recorded for 30
min. Any pressure changes in the film usually
occurred during the first few minutes of a run with
little if any change noted towards the end of the
period. A series of such runs at varying protein
concentration established a pressure rise-protein
concentration profile for that particular lipid
monolayer/subphase combination. These profiles were
usually in the shape of a sharply rising pressure at
low protein concentration followed by a plateau at
higher concentration. The knees of the plateaus
generally occurred between 2-3 mg/L protein concentration.

(3)   The pressure rise in a lipid monolayer observed upon
injecting protein into the subphase was recorded for
a series of lipid monolayers with different initial
pressures. Sufficient protein was injected into the
subphase to give a protein concentration of 5 mg/L,
which was well into the plateau region established
in experiment (2) above. These experiments gave the
pressure rise in the lipid film (upon injection of
protein into the subphase) as a function of initial
lipid film pressure (just before injection of the
protein). This allowed us to estimate a "critical"
initial lipid pressure, above which no change in the
monolayer pressure was observed upon injection of
protein into the subphase.

(4)   In all subsequent experiments, the initial lipid
pressure was kept above the "critical pressure" to
prevent the protein molecule from inserting a
portion of itself into the monolayer. The effect of
solution (subphase) pH and calcium concentration on
lipid-protein binding in the monolayers was studied
by spreading lipid onto subphases of different
composition, injecting the protein and then trans-
ferring the film to quartz plates for UV and CD
analysis. Eight quartz plates were partially im-
mersed in the subphase prior to spreading the lipid.
After spreading the lipid, injecting the protein,
and equilibrating the film for about 30 min., the
monolayers were transferred to the quartz plates (at
constant film pressure) and mounted in the UV or CD
instrument for spectroscopic analysis. The spectral
range was 180-350 nm for UV spectroscopy and 180-260
nm for CD spectropolarimetry. Control experiments
showed that only proteins contributed significantly
to either the UV or the CD spectra over the ranges
studied. Additional details about equipment and
methods for CD (5) and UV (7) spectroscopy have been
given.

Adsorption of the proteins to the quartz plates used for
film transfer and to the zwitterionic choline was antici-
pated. Since we were interested in adsorption only to
the charged lipid POPG, control experiments were run and
contributions to the UV spectra from protein adsorbed to
the quartz plates and POPC was subtracted from the UV
absorbance. For every UV experiment using POPC/POPG
(65/35 mol%) there was a control run using identical
procedures (subphase pH, protein concentration etc.)
except for the lipid film which was a monolayer of POPC.
In the case of calcium free systems, the UV absorption
observed with the control films was about 20-30% of that
observed with monolayers containing POPG; this contribution was subtracted before calculating the POPG-protein
binding. Light scattering was noticeable in films transferred from subphases containing 10 mM sodium chloride;
we employed a log-log extrapolation procedure to correct
for scattering contribution to the 190 nm protein absorption band (8). Contributions from this source were estimated to be about 20-30% of the total UV absorption
signal observed at 190 nm with 10 mM sodium chloride in
the subphase.

Results and Discussion

The results discussed here were first published separately for β-LG (6) and for α-LA and BSA (Cornell, D. G.;
Patterson, D. L.; Hoban, N.; Colloid Interface Sci.,
1990, in press.) Together these two reports give a
picture of a lipid-protein-calcium interaction scheme
which follows mass action principles and involves an
electrostatic mechanism. Of particular importance are
the electrical properties of the proteins in relation to
the solution pH as discussed below.
    The purpose of steps 1-3 above was to establish
conditions under which we were confident that the proteins were interacting with only the head groups of the
lipids while at the same time achieving maximum lipid-
protein complexation. Step 1 established the maximum
surface pressure the various proteins could support at
the air/water interface over the concentration range
studied. These results are shown in Figure 2. In step
2, where protein was injected beneath a lipid monolayer,
the initial lipid film pressure was higher than the
pressure achieved by proteins at the air/water interface.
This effectively suppressed the formation of patches of
pure proteins in a monolayer of lipid. Any film pressure
rise observed upon injecting protein into the subphase
was taken as evidence of insertion of a portion of the
protein molecule into the lipid monolayer in an intimate
lipid protein mixture. The results are shown in Figure
3. In this step we were also able to establish the

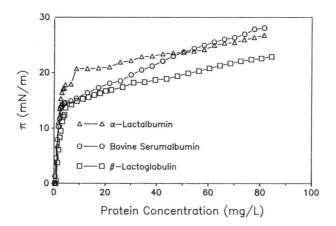

Figure 2. The surface pressure (surface tension
lowering) of proteins in Walpole's acetate pH 4.4 as a
function of concentration.

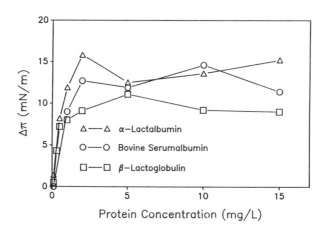

Figure 3. The rise in surface pressure of a phospholipid (POPC/POPG, 65/35 mol%) monolayer upon injection
of protein into the underlying subphase. The initial
pressure of the lipid films was 22 mN/m. Subphase;
Walpole's acetate pH 4.4.

"saturation" concentration for the proteins; the concentration (or narrow range of concentrations) above which no further evidence of lipid-protein interaction was observed. Despite the variability in the pressure rise shown by the three proteins in Figure 3, "saturation" concentration appears to occur between 2 to 5 mg/L protein concentration in the subphase for all proteins. In all subsequent experiments, sufficient protein was injected into the subphase to achieve a concentration of 5mg/L.

The magnitude of the pressure rise observed in the lipid monolayer upon injection of protein into the subphase depended upon the value of the initial monolayer pressure. When the initial lipid pressure was above a certain "critical" value, little or no change in the pressure of the lipid monolayer was observed upon injection of protein into the subphase. This was taken to mean that insertion of the protein into the monolayer was effectively suppressed. The "critical pressure" was estimated by extrapolating a plot of pressure rise vs. initial pressure to zero pressure rise as shown in Figure 4. While the penetration of protein into the lipid portion of the oil/water interface is certainly an interesting mechanism for the stabilization of emulsions, we wished to study electrostatic interactions at the oil/water interface without the complications of protein insertion. For all subsequent experiments reported here 35 mN/m was the initial lipid pressure. In all these cases the change in the film pressure (±) after injection of the protein was less than 1/2 mN/m for the duration of each run, about 30-45 min.

The amount of protein interacting with the phospholipids in monolayers at various subphase pH's was determined from the UV spectra of the transferred monolayers using what is essentially Beer's Law for films, namely:

$$A = n\epsilon_M \Gamma / 6.02 \times 10^{20} \tag{1}$$

where A is the absorbance of n monolayers, ϵ_M is the molar absorptivity of the protein, and Γ is the number of protein molecules per square centimeter in a single monolayer. The factor $n\Gamma/6.02 \times 10^{20}$ has the same dimensionality (mass per squared length) and numerical value as the lc product (path length in centimeters times concentration in moles per liter) of the classical Beer's law for solutions. A straightforward calculation then gave the surface concentration in milligrams of protein per square meter of monolayer as presented in Table I.

The results in Table I clearly show the effect of pH on protein binding to the monolayer lipids and supports an electrostatic mechanism for lipid-protein interaction. At acid pH's the proteins carry a significant positive charge over their surface which can bind to the negatively charged lipid. At neutral pH where both the lipids

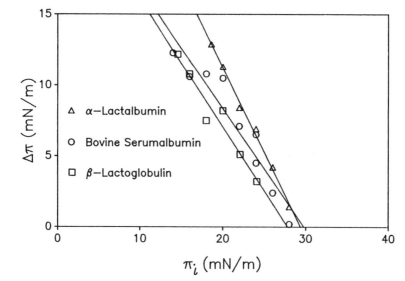

Figure 4. The increase in lipid monolayer pressure upon injecting protein into the subphase as a function of the initial film pressure. Protein concentration in the pH 4.4 subphase, 5 mg/L.

Table I. Protein Concentration in
Phospholipid-Protein Monolayers

Protein		BSA	α-LA	β-LG
Subphase	Concentration Units[2]			
Tris pH 7.0	mg/m^2	0.09 [.05][1]	0	0.4 [.1]
	molec./cm^2	8^2 X 10^{10}	0	1.3 X 10^{12}
Walpole's Acetate pH 5.2	mg/m^2	1.1 [.2]	–	–
	molec./cm^2	1.0 X 10^{12}	–	–
Walpole's Acetate pH 4.4	mg/m^2	3.1 [.3]	2.6 [.6]	3.0 [.2]
	molec./cm^2	2.8 X 10^{12}	1.1 X 10^{13}	1.0 X 10^{13}
Walpole's Acetate pH 4.4 10 mM NaCl	mg/m^2	1.9 [.4]	0.9 [.3]	–
	molec./cm^2	1.7 X 10^{12}	3.7 X 10^{12}	–

[1] Numbers in brackets are standard errors of 3-4
   replicates.
[2] Errors in the molec./cm^2 units have been omitted for
   clarity but are proportionally identical to the errors
   for the corresponding mg/M^2 units.

and proteins carry a net negative charge there is electrostatic repulsion and a reduced tendency to interact.
BSA shows this tendency most clearly. At pH 7, little if
any lipid-protein complex is formed (.09 mg/M^2 is of
doubtful statistical significance). At the isoionic
point of BSA, pH 5.2, it is clear that the protein is
interacting with the lipid. At a more acid pH, 4.4, the
coverage has reached 3.1 mg/M^2. A concentration this
high suggests the formation of a contiguous film of
protein in its native (globular) state one molecular
layer thick. Values of this order have been reported for
many proteins adsorbed at the air/water interface (mi-
grating from bulk solution), where the measurement techniques were ellipsometry (9) and surface radioactivity
(10). Spread films, formed by direct application of the
protein from solution onto the surface of water, generally have lower surface concentrations than adsorbed
films at equivalent surface pressures (11). It is
thought that some degree of unfolding occurs in a protein
molecule when it is applied to the surface of water,
subject to constraints such as intramolecular disulfide

cross linking (11). In earlier work, an increase in beta
sheet was observed in ß-LG spread at the air/water interface (2). In the current work, no such change in the
secondary structure of any of the proteins was found.
The CD spectra of the proteins in the lipid-protein monolayers (Figure 5) were similar to their spectra in
solution as determined in the authors' laboratory. This
suggests that there was no gross changes in any of the
proteins secondary structure upon interacting with the
phospholipids in the monolayers. This is consistent with
the picture of globular proteins adsorbing (without unfolding) to the phospholipids in a monolayer.
   Perhaps the most interesting finding presented in
Table I is the amount of complexation that occurs with a
lipid by a protein in its zwitterionic state. This is
shown by the film concentrations for BSA at pH 5.2 and
α-LA at pH 4.4. At first sight this observation seems to
run counter to an electrostatic mechanism for lipid-
protein interaction, since a protein at its isoionic
point has a **net** neutral charge. If the charge distribution is homogeneous over the surface of such a protein,
then the molecule will appear to be neutral even at very
close range. As the three dimensional structure of more
and more proteins are determined however, it is becoming
increasingly clear that the charge distribution over the
surface of most proteins is anything but homogeneous.
Consistent with this are electrostatic field calculations
(12-15) which show that clusters of basic and acidic
residues commonly form large potential envelopes over the
protein surface. The patches of positive and negative
charges will cause significant interaction when they are
brought into close enough proximity to a charge of opposite sign. Lowering the pH of the aqueous phase below
the isoionic point of the protein will give it a net
positive charge, and the total area of the positive
patches will grow. This should facilitate docking of the
globular proteins to the negatively charged lipids, resulting in increased adsorption of the proteins at the
oil/water interface. All of this is consistent with the
results presented in Table I.
   Calcium ions can also bind to negatively charged
phospholipids and thus compete with proteins in a lipid-
protein-calcium interaction scheme. Since calcium is a
major component of milk and many other foods, knowledge
of the role it plays in the adsorption of proteins at the
oil/water interface is essential to the understanding of
emulsion stabilization in such complex systems. We have
studied the effect of calcium on lipid-protein interaction in several of the systems listed in Table I simply
by adding calcium to the subphase buffer before spreading
the lipid. Except for calcium in the buffers, all other
conditions and procedures were as in step 4 above under
methods. The amount of protein complexing with lipids in
the presence of calcium was determined from the UV spec-

tra of transferred films and formula 1 above. The
quantity of protein so obtained was compared to the
values obtained for the same system in the absence of
calcium (Table I), and plotted as the percent POPG/
protein complex vs. calcium concentration as shown in
Figure 6. The solid lines represent the best fit to the
discrete data points based on the assumption of saturation binding over the entire range of calcium concentrations studied.   In the case of BSA at pH 4.4 this
resulted in a relatively poor fit at the lowest concentrations.   Due to the variability in the data, the
significance of this "poor fit" is uncertain. Because of
this we have chosen to discuss the data only in terms of
overall trends (see below).
     When either α-LA or BSA was injected into a subphase
near the respective proteins' isoionic point, calcium
ions at the micromolar level interfered with lipid-
protein binding.   This is shown by the steep curves near
the ordinate in Figure 6 where the subphase buffer is pH
4.4 for α-LA and pH 5.2 for BSA.   These results imply
that the apparent dissociation constant of the lipid-
calcium complex is of the order of micromoles or less.
When the pH of the subphase was below the isoionic point
of the protein, millimolar levels of calcium were required to suppress the lipid-protein interaction.   This
is shown by the upper curves in Figure 6 where the subphase buffer is pH 4.4 for both ß-LG and BSA.   These
results in turn imply that when the protein carries a net
positive charge, it is better able to compete with the
calcium ions for complexation to the lipid.
     Assuming that calcium binds to POPG in a 1/1 complex
as found by Lau et al. (16), we can summarize the binding
of protein and calcium to monolayer lipid by Equation 2,

$$P^{z+} + nCaL^+ \rightleftharpoons PL_n^{(z-n)+} + nCa^{2+} \qquad (2)$$

where P is protein carrying charge z, Ca is calcium, L is
the lipid POPG⁻, and n represents the number of POPG
lipids bound per protein molecule.   The protein P^{z+} and
calcium Ca^{2+} are in solution; the protein-lipid and
calcium-lipid complexes $PL^{(z-n)+}$ and CaL^+ are components of
the monolayer.   The formalism for expressing the interaction between species in solution and components of a
monolayer is essentially the same as that for the mass
action law in solution (16).
     Clearly what is being proposed here is an electro-
static mechanism for calcium-protein-lipid interactions
in monolayers.   If this mechanism is correct, then salt
should have an effect on lipid-protein binding.   This is
borne out by the results with 10 mM NaCl in the subphase
as shown in Table I.   Binding of both α-LA and BSA is
slightly reduced in this media compared to the corresponding salt free system.

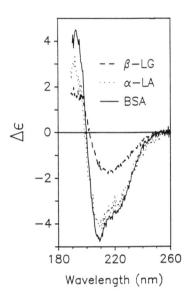

Figure 5. The circular dichroism spectra of lipid-
protein monolayers; sixteen monolayers for each
lipid-protein set. The proteins were adsorbed from pH
4.4 Walpole's acetate onto POPC/POPG monolayers.

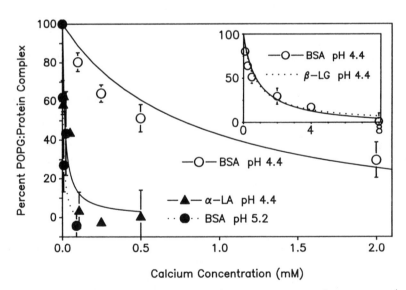

Figure 6. The effect of calcium concentration and pH
on the adsorption of proteins to POPG in monolayers.

If expression 2 is correct one would expect that the equilibrium could be driven to the right, in the direction of increased protein-lipid binding and reduced calcium-lipid binding, simply by increasing the protein concentration in the subphase. A small shift of this sort was found with ß-LG. Increasing the protein concentration by a factor of 10 (to 50 mg/L) at a calcium concentration of 2 mM increased the amount of lipid-protein complex observed from 25% to about 36% (6). Assuming that equilibrium 2 is for n identical noninteracting binding sites, it can be shown by a simple mass action expression that this relative insensitivity to protein concentration implies that n is "large". The assumption of n identical and independent binding sites is not generally applicable however, and the determination of multiple equilibria constants can be rather involved (17, 18). Furthermore, the scatter in the data of Figure 6 is too great to allow an analysis in terms of multiple equilibria. The protein concentration used in most of this work was quite low (5 mg/L or 0.0005%); increasing this value to levels more typical for many foods (3 g/L for ß-LG in milk for example (19)) should increase protein binding to charged lipids.

Conclusion

The colloid stability of foods depends on the complex interplay of many factors. Electrostatic as well as hydrophobic interactions are important stabilizing influences. The aqueous milieu, its pH, mineral content, etc., help determine the electrical forces between surfaces which play such a vital role in colloid stability. In this work we have shown how the interrelationship between the pH of a solution and the isoionic pH of a protein determines its ability to compete with calcium for adsorption onto negatively charged phospholipid surfaces. This suggests that the electrical properties of the individual proteins are among the many factors that must be considered when the development of a new food is being contemplated.

Literature Cited

1. Keenan, T. W.; Dylewski, D. P.; Woodford, T. A.; Ford, R. H. In Developments in Dairy Chemistry; Fox, P. F., Ed.; Applied Science: New York, 1982; Vol. 2, p 83.
2. Cornell, D. G. J. Colloid Interface Sci., 1982, 88, 536.
3. Sober, H. A. Handbook of Biochemistry; 2nd. Ed; The Chemical Rubber Company: Cleveland, OH, 1970; p J234.
4. Gaines, G. L. Insoluble Monolayers at Liquid-Gas Interfaces; Wiley: New York, 1966.

5.   Cornell, D. G.   J. Colloid Interface Sci., 1979, 70,
     167.
6.   Cornell, D. G.; Patterson, D. L.   J. Ag. Food Chem.,
     1989, 37, 1455.
7.   Cornell, D. G.   J. Colloid Interface Sci., 1984, 98,
     283.
8.   Wetlaufer, D. B., (1962) "Ultraviolet Spectra of
     Proteins and Amino Acids" in Advances in Protein
     Chemistry; Vol 17. Anfinsen, C. B.; Anson M. L.;
     Bailey, K.; Edsall, J. T., Eds.; Academic Press: NY;
     pp 304-390.
9.   Benjamins, J.; de Feijter, J. A.; Evans, M. T. A.;
     Graham, D. E.; Phillips, M. C. Faraday Discuss.
     Chem. Soc. 1975, 59, 218.
10. Graham, D. E.; Phillips, M. C. J. Colloid Interface
     Sci. 1979, 70, 415.
11. Mitchell, J.; Irons, L.; Palmer, G. J. Biochim.
     Biophys. Acta 1970, 200, 138.
12. Warwicker, J.; Watson, G. C.   J. Mol. Biol. 1982,
     157, 671.
13. Rogers, N. K.; Sternberg, M. J. E. J. Mol. Biol.
     1984, 174, 527.
14. Klapper, I.; Hagstrom, R.; Fine, R.; Sharp, K.;
     Honig, B. Proteins 1986, 1, 47.
15. Warwicker, J. J. Theoret. Biol. 1986, 121, 199.
16. Lau, A.; McLaughlin A.; McLaughlin, S. Biochim.
     Biophys Acta 1981, 645, 279.
17. Tanford, C. Physical Chemistry of Macromolecules;
     Wiley: New York, 1961; pp 526-586.
18. Steinhardt, J.; Reynolds, J. A. Multiple Equilibria
     in Proteins; Academic: New York, 1969; pp 217-231.
19. Gordon, W. G.; Kalan, E. B. in Fundamentals of Dairy
     Chemistry; 2nd Ed. Webb, B. H., Johnson, A. H.,
     Alford, J. A., Eds.; AVI: New York, 1974; p 113.

RECEIVED October 18, 1990

Chapter 10

Interactions Between Milk Proteins and Lipids

A Mobility Study

M. Le Meste, B. Closs, J. L. Courthaudon, and B. Colas

Ecole Nationale Superieure de Biologie Appliquee a la Nutrition et a l'Alimentation, Campus Universitaire de Montmuzard, 21000 Dijon, France

The interactions between the main milk proteins (whey proteins and caseins) and nitroxide homologs of fatty acids, either dispersed in a solution or included in oil droplets, were studied using Electron Spin Resonance. The affinity of the proteins for the fatty acids increased in the order : α-lactalbumin $< \alpha_{s1}$, β and whole casein $< \beta$-lactoglobulin and whey proteins. It appeared that the interactions between milk proteins and lipids occured through their polar groups. Chemical and biochemical modifications of the amino residues of the proteins confirmed the contribution of these residues to the interaction. No evidence of any significant modification of the conformation of the proteins or of the organization of the lipid monolayer at the interface between water and the oil droplets was observed.

Many of the oil in water emulsions produced by the food industry are stabilized by milk proteins. The main structural characteristics of these proteins are relatively well known. However the mechanism of their action for emulsion stabilization remains only partially understood. Among the factors which are presumed to intervene in this mechanism, we selected to study more particularly two of them : 1) the affinity of the milk proteins for the lipids. The real importance of this parameter is not known , nor is the nature of the protein-lipid interaction. 2) the contribution of the flexible side chains of the proteins. Recent studies have stressed the essential role played by the more flexible regions of macromolecules in the establishment of interactions with other molecules, either small solutes or polymers. Lysines are among the most flexible and accessible side chains of proteins. It can be mentioned for example that these residues are frequently found in catalytic and allosteric sites as well as on the protein surface where they may enhance solubility or provide a site for the electrostatic attachment of peripheral membrane proteins to phospholipids (1).

The establishment of interactions between proteins and lipids can be deduced either from a change in the mobility of lipidic molecules dispersed in a solution or included in oil droplets, or from a modification of the motions of the interacting residues of the proteins upon emulsification. Thus, we based our approach on mobility measurements using the Electron Spin Resonance (ESR) method.

0097–6156/91/0454–0137$06.00/0
© 1991 American Chemical Society

ESR is a powerful technique widely used to derive structural and dynamic informations about biological membranes and enzymes. Only paramagnetic species, i.e. molecules with an unpaired electron, are detected with this technique. We used stable nitroxide radicals, either paramagnetic homologs of fatty acids or isothiocyanate nitroxides, as spin-labels covalently bound on the amino groups of the proteins. Two homologs of fatty acids were selected, the 5SA with the nitroxide moiety close to the polar head of the fatty acid and reflecting the behaviour of this polar head, and the 16 SA with the nitroxide moiety close to the CH_3 end and reflecting the behaviour of the apolar chain (Figure 1). Changes in the mobility of spin-labelled residues of proteins have been measured in order to reveal either interactions with the lipid phase or modifications of the protein conformation induced by the chemical or physical treatments performed on milk proteins in order to modify the accessibility of their polar groups.

In the present work, on the one hand, we compared the relative affinities of the main milk proteins for the lipids; on the other hand, we assessed the contribution of the flexible side chains of the proteins to this interaction.

Materials and Methods

Protein Sample.   The control, whole casein was prepared in the laboratory from fresh skimmed milk by isoelectric precipitation at pH 4.6, washed three times with distilled water then dissolved at pH 7. After two additional precipitations, caseinate solutions were freeze-dried. The solution remaining after casein precipitation was filtered, dialysed at 4°C against distilled water, concentrated on a Millipore ultrafiltration unit, then freeze-dried (this preparation will be called whey proteins in the text). Individual caseins were purified by ion exchange chromatography on DEAE Trisacryl M using Tris HCl buffer 5mM, pH 8, urea 4.5M, dithiotreitol 6.4 x 10^{-5} M, NaCl concentration from 0.2 to 0.4M. α-Lactalbumin and β-lactoglobulin were also purified from the whey protein preparation by ion exchange chromatography on DEAE Trisacryl M using Tris HCl buffer 0.02M, pH 7.6, NaCl concentration from 0 to 0.3 M. Slab polyacrylamide electrophoresis confirmed the purity of the samples.

The protein content of the samples was estimated by the Lowry method (2) using bovine serum albumin (BSA) as a standard.

Trypsic Hydrolysis of β-Lactoglobulin.   The protein was digested with TPCK trypsin (Sigma) at an enzyme/protein ratio of 1/100 (w/w), at 37°C, while maintaining the pH at 7.5 with a pH stat (Tacussel). The enzyme reaction was stopped by addition of soya trypsic inhibitor (type IS-Sigma) at an inhibitor/enzyme ratio of 1/1 (w/w).

Disulfide Bridge Reduction and SH Carboxymethylation.   The whey protein S-S bridges were reduced using the method of Konisberg (3), at 50°C, with use of dithiothreitol (54.5 mM for a 1% protein solution) ,in buffer TRIS HCl 0.5 M with urea 8 M and EDTA 2 mM. Free SH content was estimated using DTNB (5,5'-dithiobis 2-nitrobenzoic acid) 0.1 M of Iodoacetamide was used for blocking the released free SH, then the solution was dialysed against distilled water and freeze-dried.

Glycosylation of Casein.   Carbohydrate modified caseins were obtained using the method of Lee et al (4). The degree of modification was determined from the amount of unreacted ε-amino groups of lysine using the TNBS method (TNBS : 2-4-6- trinitrobenzene sulphonic acid) (5). It was expressed as the percentage of lysine residues which had been modified.

Emulsions.__ One volume of soya oil was added to 3 volumes of a 2.5% (w/v) protein solution at pH 7. The mixture was homogenized with a Polytron PCU-2 at 12000 rpm for 2 min.

ESR Analysis.-- An important property of the nitroxides is the fact that their ESR spectra are very sensitive to their mobility. This results from the anisotropy of its magnetic parameters. Different spectra are obtained according to the nitroxide axis being oriented parallel or perpendicular to the external magnetic field.They differ mainly by the position of the lines (g factor) and by the distance between these lines (hyperfine coupling constant a_o). In dilute solutions, as a consequence of the fast molecular reorientation, there is an averaging process. The g and a_o parameters are determined by the average values of the g and A tensors. A powder spectrum is obtained when the nitroxides are randomly oriented and have very slow motions. The spectrum is the sum of the spectra corresponding to all possible orientations for the nitroxides. Intermediate spectra or a superposition of both previous kinds are also often obtained. Changes in the relative contribution of both previous types of spectra are reflected by modifications in the line amplitude ratio $R = I/M$ (Figure 2) ; I corresponding to radicals with slow motions and M to radicals with rapid motions. Despite the fact that it does not represent quantitatively the proportion of both populations we used this line amplitude ratio to express changes in the proportion of fatty acid interacting strongly with the proteins. This ratio is very sensitive to small changes in this proportion. It was considered as representing, qualitatively, the affinity of the protein for the fatty acids.

The mobility can be expressed by two parameters, the rotational correlation time τ_c or the order parameter S. The parameter τ_c is used for unoriented and isotropic systems. It is calculated from the features of the ESR spectra (line width and amplitudes). Drot is inversely proportional to τ_c , Drot.$= 1/(6 \ \tau_c)$; it represents the reorientation frequency of the molecule. ESR is sensitive to changes in mobility in the range of correlation times from 10^{-7} to 10^{-11} s. The order parameter S is used for molecules which have an anisotropic motion because of their structure or of their environment . S is calculated from the coupling constants. It describes the average orientation of the nitroxide. For the fatty acid homologs 5SA and 16SA motions occur preferentially around the axis of the hydrocarbon chain which coincides with the spin label Z axis. However for rapid motions, and if there is no preferred orientation in the system, the overall behaviour can be considered as isotropic and the mobility can be expressed by τ_c or Drot.. The latter parameter was used to represent the mobility of the fatty acids remaining in the solution. The coupling constant a_o is sensitive to the polarity of the medium surrounding these nitroxides, a_o = 14G for an apolar medium, a_o = 17G for a polar environment.

The paramagnetic homologs of fatty acids were dissolved in acetone, then the solvent was evaporated in order to form a film of fatty acid on the walls of the vials; the protein solution was added and mixed, finally the ESR analysis was performed either on this solution or after addition of oil and emulsification. The molecular ratio fatty acid / protein was 0.5. For protein spin-labelling, 2mg of isothiocyanato-nitroxide was incubated with 10ml of a 2% (w/v) protein solution for 24h at pH 9 and 4°C. Then the solution was dialysed against distilled water at 4°C for 48h. On the average less than 1 nitroxide was covalently bound on each protein molecule.

ESR experiments were performed at room temperature (20 ± 1°C) with a Varian E9 spectrometer using an aqueous cell. The ESR conditions were : microwave power : 10mW, modulation amplitude : 0.5G . The spin labels 5SA, 16SA and the isothiocyanato-nitroxide were from Aldrich.

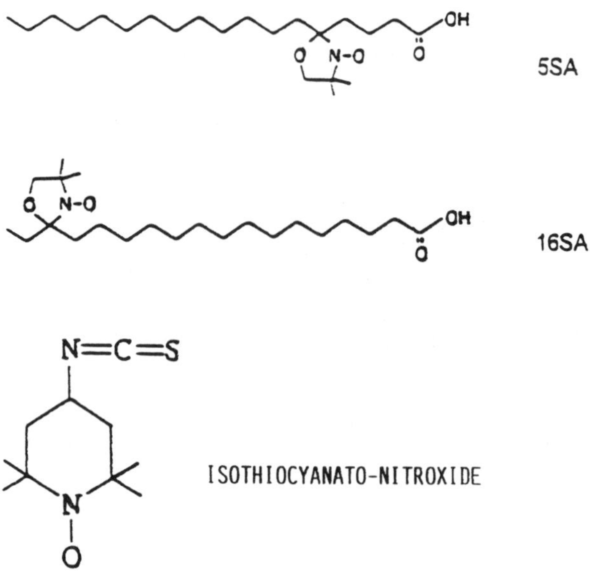

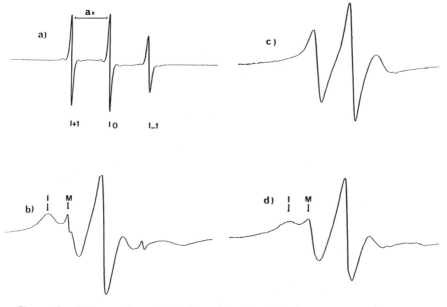

Figure 1 : Nitroxide radicals

Figure 2 : ESR spectra of 5SA in : a) buffer ; b) whey protein solution ; c) oil/buffer emulsion ; d) previous emulsion stabilized by whey proteins. I : I+1 line of the spectrum representing the nitroxides with slow motions ; M : I+1 line of the spectrum representing the mobile nitroxides.

Results and Discussion

The interactions of the milk proteins with the 5SA and the 16SA were studied in parallel. It was assumed that the affinity of the proteins for both fatty acids were similar but that the consequences of these interactions on the mobility of the nitroxides differed depending upon their location on the chain: close to the polar head or at the apolar end. The mean relative errors for ESR measurements were: Drot.: + 5%; I/M: + 10%; a_0 : + 0.25G.

Mobility of the Polar End of the Fatty Acid (5SA Nitroxide) (Table I). The spectra obtained with 5SA in buffer alone or in presence of whey proteins are shown in Figure 2 a-b. The spectrum of 5SA (0.25mg per ml) in buffer reveals that the nitroxides had rapid motions and were well dispersed. No dipolar interaction which should have resulted from a colloidal dispersion was observed. When 25mg of whey proteins were added to 1ml of the previous solution, a high proportion of fatty acids lost mobility. The ESR spectrum was composed of two components reflecting two populations of radicals: one with slow motions, corresponding to the powder spectrum and with a Drot. $< 10^{-6}$ s (population I), the other one with much more rapid motions (population M) , its mobility being similar to that in buffer without protein. The comparison with simulated spectra allows us to estimate that more than 95% of the radicals had slow motions. They were assumed to interact with the proteins. Results concerning the main milk proteins are given in Table I.

Table I. Interactions between milk proteins and 5 doxyl stearic acid (5SA) in solution (buffer 10 mM, pH 7, protein concentration: 2.5% (w/v), molecular ratio fatty acid / protein: 0.5)

Sample	R = I/M	D_{rot} (M) s^{-1}	a_0 (M) G
Buffer	0	$1.0 \ 10^9$	16.4
α lactalbumin	0.3	$0.8 \ 10^9$	15.7
R^1-α lactalbumin	0.6	$0.8 \ 10^9$	15.7
β lactoglobulin	1.0	$0.5 \ 10^9$	15.7
R^1-β lactoglobulin	1.4		
Whey proteins	1.0	$0.5 \ 10^9$	15.7
α_s casein	0.4	$0.8 \ 10^9$	15.7
70% G^2 — α_s casein	0.4	$0.8 \ 10^9$	15.7
β casein	0.4	$0.7 \ 10^9$	15.8
84% G^2-β casein	0.1	$0.8 \ 10^9$	15.8
Whole caseinate	0.4	$0.7 \ 10^9$	15.8
84% G^2-caseinate	0.1	$0.7 \ 10^9$	15.8

1. After reduction of the disulfide bridges and carboxymethylation of the SH groups.
2. glycosylated caseins.

The proteins can be ranged in order of increasing values for R ,i.e. of increasing affinity for the fatty acids:

α-lactalbumin$<$ caseins $<$ β-lactoglobulin $=$ whey proteins

The spectrum obtained with α-lactalbumin reveals a lower affinity of this protein for the fatty acids compared to the other milk proteins. The α and β fractions of casein and the whole caseinate presented the same affinity for the 5SA (approximately 75% of the fatty acids initially dispersed in the solution had slow motions). The spectra obtained with whey proteins and β-lactoglobulin, its major component, were similar. The high affinity of β-lactoglobulin for the fatty acids was confirmed at pH 5;, the pH at which this protein protected the fatty acids against aggregation whereas α-lactalbumin did not. The mobility of the fatty acids remaining in solution (Drot.(M)) was similar with all the proteins and not very different from their mobility in buffer. The coupling constant, a_o, reflects the polarity of the surrounding medium, decreased slightly when proteins were added to the buffer.

We estimated for our conditions of pH and concentration, and in the presence of 1 mole of β-lactoglobulin, that the mobility of on average 0.5 mole of fatty acid was reduced. It was previously demonstrated that one molecule of this protein could bind approximately 1 apolar molecule (6). The good binding properties of β-lactoglobulin with apolar substances is well known (7 - 10). A similarity in the structures of β-lactoglobulin and of the Plasma retinol binding protein was suggested by Papiz et al (10).

On the basis of electrophoretic mobility measurements and from the time dependant changes in composition at the oil/water interface, Dickinson et al. (11) measured the exchange of proteins between the interface and the bulk solution. The authors observed differences in the behaviours of caseins and whey proteins : there is much more exchange at the interface between the casein monomers (α and β) than between the whey proteins , β-lactoglobulin was more difficult to displace from the interface than α-lactalbumin. Similarly, using electrophoresis, Closs (12) demonstrated with mixtures of the main components of the whey proteins, that β-lactoglobulin was preferentially associated with the oil phase. These data are in agreement with the ESR results and suggest that the interactions between milk proteins and lipids are reversible.

Whey proteins are globular proteins whereas caseins are mainly random coils. Whey proteins are rigid because of the presence of organized structures and of disulfide bridges; caseins are very flexible molecules. Thus it seems that the overall conformation of the proteins and their overall flexibility are not the determinant factors for their ability to interact with fatty acids. However, α-lactalbumin which contains 4 disulfide bridges, is particularly rigid, dense and resistant to enzymatic hydrolysis. The strongly reduced accessibility of all its residues could explain its relatively low affinity for the fatty acids. The total hydrophobicity of these proteins are quite similar but the accessibility of hydrophobic residues is much higher for caseinates than for the whey proteins. Again, the affinity for fatty acids could not be explained by this characteristic. However it must be mentioned that the response of globular and random coil proteins may differ upon changing the fatty acid / protein ratio.

Mobility of the Apolar Chain (16SA) (Table II). In buffer, the mobility of 16SA was of the same order as the mobility measured for the 5SA. In presence of whey proteins the proportion of radicals with reduced mobility was close to 0 for 16SA. The same results were obtained with caseins. The mobility of these nitroxides was not significantly modified by the presence of the proteins. It appears from these results that the interactions between milk proteins and fatty acid occurred preferentially through the polar head of the fatty acid.

Behaviour of the Fatty Acids Included into the Triglyceridic Droplets (Table II).
The nitroxide fatty acids were also used as probes dispersed in the oil, in order to

study the interactions between milk proteins and triglycerides in food emulsions. Upon incorporation into the oil droplets, the mobility of the nitroxide was reduced by a factor of approximately 10, apart from the position of the nitroxide on the chain of the fatty acid. The apolar nature (i.e. lipidic) of the environment of the nitroxides was confirmed by the a_o value. The decomposition of the I-1 line indicates some partition between the aqueous and lipidic phases (Figure 2c). Table II shows that, in presence of whey proteins, strongly immobilized 5SA were observed, whereas 16SA included within the fat droplets was less sensitive to the presence of whey proteins; no population with slow motions was apparent. The differences between the behaviours of β-lactoglobulin and α-lactalbumin, already observed in solutions, were confirmed with emulsions containing the 5SA nitroxides.

Table II. Interactions between whey proteins and 5-and 16-doxyl stearic acids (5SA; 16SA) in solutions and in emulsions

Sample	I/M	D_{rot} (M) s^{-1}	a_0 (M) G
5SA Nitroxide			
Buffer	0	$1.0 \ 10^9$	16.4
Whey proteins	1.0	$0.5 \ 10^9$	15.7
Emulsion (buffer)	0	$1.0 \ 10^8$	14.5
Emulsion + whey prot.	0.8	$7.0 \ 10^7$	14.5
16SA Nitroxide			
Buffer	0	$1.6 \ 10^9$	15.7
Whey proteins	0	$0.9 \ 10^9$	15.5
Emulsion (buffer)	0	$1.6 \ 10^8$	14.5
Emulsion + whey prot.	0	$2.3 \ 10^8$	14.5

Concerning the fatty acids remaining in the lipid phase and less affected by the presence of the proteins (population M), only slight changes in mobility could be observed. This may reflect limited modifications in the organization of the lipids : a tendency to a reduction in the mobility of the polar head and on the opposite end to an increase in the mobility of apolar chains.

Triglycerides belong to the class of polar lipids which has surface solubility (13). At the triglyceride/ water interface the typical liquid cristal structure is obtained. Triglycerides constitute a monolayer with their polar heads (glycerol moiety and carbonyl groups) ordered and in contact with the solvent and with their apolar chain relatively free to move and in contact with the oil phase (13). Fatty acids, being more polar than the triglycerides, may concentrate at the oil/water interface. Moreover some exchange with the aqueous phase may occur and the fatty acids interacting with the protein could be either in the solution or at the surface of the oil droplets. In emulsions as in solution, 16SA was less sensitive than 5SA to the presence of the proteins. Thus, no evidence of any penetration of the protein inside the lipid monolayer could be deduced from our results. Whey proteins seem to bind to lipids of soya oil as extrinsic membrane proteins do to phospholipids.

Motions of the Protein Side Chains.   This experiment was performed with whey proteins and with its purified components. With β-lactoglobulin in dilute solutions (Figure 3), two populations of nitroxides can be distinguished. The sharp components (M) in the corresponding spectra, indicating high mobility of the spin-labelled groups, can be ascribed to surface residues. The spin-labelled residues with constrained motions may be trapped within the organized regions of the proteins or within aggregates. Both populations were also observed with whey proteins (Table III). On the other hand, only one population of fast moving spin labelled residues was apparent with α-lactalbumin (Table III) probably because the very high density of this protein inhibited the labelling of buried amino groups.

Table III. Effect of emulsification on the characteristics of the ESR spectra of the spin-labelled amino residues of whey proteins

Sample	I/M	D_{rot} (M)	a_0 (M)
in solution			
Whey proteins	0.3	$1.1\ 10^9$	16.9
α lactalbumin	≈ 0	$2.6\ 10^9$	17.1
β lactoglobulin	0.2	$0.8\ 10^9$	16.8
in emulsion			
Whey proteins	0.3	$0.2\ 10^9$	15.6
α lactalbumin	0	$0.3\ 10^9$	15.4
β lactoglobulin	0.3	$0.2\ 10^9$	15.5

The dynamic of lysine residues in proteins is a function of the level of structure in the protein. The accessibility and mobility of the lysine residues in β-lactoglobulin was studied by Brown et al. (1) ; β-lactoglobulin appeared to have several populations of lysine in distinctly different environments. In aqueous buffer at pH 7.5, the equivalent of six to seven residues (Lys. 8,14, 60,77,100,101) are located at the outer surface of the molecule with little restriction to side chain mobility. Lysine 135,138, 141, along the amphipatic helix, would have a correlation time similar to that of the entire molecule.The mobility of lysine 69, 70,75 may be restricted by the presence of the disulfide bonds. The spin-labelling results reflect this repartition of the lysine residues.

In   emulsions, the rotational mobility of the more mobile side chain was reduced (Table III). This effect was more pronounced for α-lactalbumin. The coupling constant indicated that the polarity of the environment of these residues was reduced and became characteristic of a less polar environment due to the presence either of neighbouring proteins or of the polar heads of the triglycerides. The proportion of spin-labelled residues with slow motions was not significantly affected by the interaction of the proteins with the oil phase. Thus, the overall conformation of the whey proteins, purified or not, did not appear to be modified significantly, when interactions with lipids occurred. Similarly Cornell and Patterson (14) observed no detectable difference between the conformations of β-lactoglobulin in solution and in the presence of a phospholipid monolayer.

Modifications of the Amino Residues of the Milk Proteins.   Papiz et al. (10) and Brown et al. (1) suggested that polar residues could participate in the interaction between the Plasma retinol binding protein and retinol. Hidalgo and Kinsella (6) observed that the interaction of β-lactoglobulin with linoleic acid 13-hydroperoxide

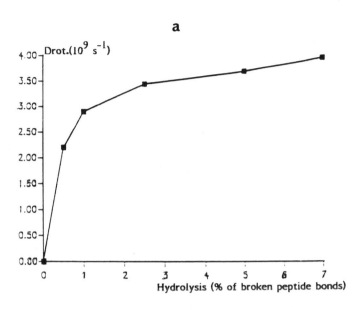

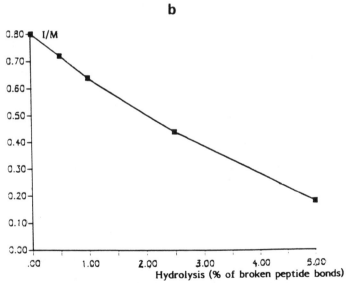

Figure 3 : Influence of a limited trypsic hydrolysis of β-lactoglobulin on (a) the rotational diffusivity of the spin-labelled amino residues of the protein; (b) the protein/fatty acid interactions as reflected by the I/M ratio.

induced a decrease in free aminogroups and that the blockage of protein amino-groups prevented the deteriorative changes which accompanied this interaction. Cornell and Patterson (14) concluded that electrostatic attractions between β-lactoglobulin and phospholipids are to be expected. In order to confirm the eventual participation of the flexible polar side chains of the milk proteins to the interaction with lipids, chemical and biochemical modifications affecting these residues were performed.

- Trypsic Hydrolysis of the β-Lactoglobulin.   Trypsin catalyzes the cleavage of peptide bonds close to lysine and glutamic acid. It can be assumed that the more accessible peptide bonds were hydrolyzed preferentially. Because of the high resistance of α-lactalbumin to enzymatic hydrolysis, the experiment was performed with β-lactoglobulin only. A spin-labelling experiment, isothiocyanate nitroxides covalently bound on the amino groups, was performed after the enzymatic hydrolysis. It revealed the opening of the protein structure. As shown on Figure 4a, the rotational diffusivity of the spin-labelled side chains increased with the progress of the hydrolysis. Despite the resulting higher accessibility of the protein residues, the fatty acid/protein interactions were inhibited by the progress of the hydrolysis (Figure 4b), i.e. probably by the elimination of the most accessible amino residues.

- Disulfide Bridges Reduction and Carboxymethylation.   As the flexibility of whey proteins is limited by disulfide bridges, a total reduction of these bridges was performed, followed by a carboxymethylation with iodoacetamide in order to prevent new bridges to be formed. Spin-labelling experiments revealed an opening of the structure of α-lactalbumin and β-lactoglobulin. The fatty acid/protein inter-actions were significantly improved by this modification (Table I). This result could be attributed to the higher number of accessible amino groups.

- Blockage of the Lysine Residues of Caseins by Glycosylation.   Caseins are random coil proteins with accessible lysines (in average 12 per monomer). We demonstrated previously that the main consequences of glycosylation on caseins were a better swelling of the modified proteins and a more dissociated state (Le Meste M., Colas B., Simatos D., Closs B., Courthaudon J.L. and Lorient D. J. Food Sci. (in press) and (15). Concerning the interactions with the fatty acids (Table I), glycosylation induced a strong lowering of the affinity of the protein for the lipids except for α_{s1} casein. It must be mentioned that more free amino residues remain in the α_{s1} casein after this treatment (5 compared to 2 for β casein).

All these experiments are in agreement with the assumption that the accessible amino residues of the milk proteins play an essential role in their interactions with lipids.

Conclusion

Even if much work remain to be done in order to confirm these preliminary results and to complete our understanding of the food protein interactions in emulsions, conclusions can already be cut : 1)milk proteins and lipids (fatty acids or triglycerides) interact preferentially through their polar groups ; 2)the polar and very accessible amino residues of the protein participate in the interaction; 3) the milk proteins can be depicted as sitting on the fat droplets without affecting the major features of the lipid organization; 4) the overall conformation of the milk proteins does not appear to be significantly modified by emulsification.

The role of the affinity of the proteins for the lipids in the emusifying properties of the proteins is not clear. The consequences of the treatments previously described on the emulsifying properties of milk properties have been studied (12). No close correlation between this fonctional property and the affinity of the protein

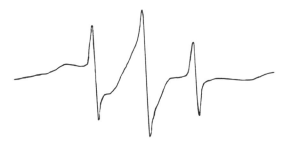

Figure 4 : ESR spectrum of the spin-labelled amino residues of β-lactoglobulin in solution (pH 7).

for the lipids was found. This could be explained by the complexity of the mechanims involved. It appeared that the affinity of the proteins for the lipids could be related to the efficiency of the proteins in lowering the interfacial tension in the emulsions.

Literature Cited

1. Brown,E.M.; Pfeffer P.E.; Kumosinsky T.F.; Greenberg R. Biochemistry, 1988, 27,5601-5610.

2. Lowry O.H.; Rosebrough N.J.; Farr A.L.; Randall R.J. J. Biol. Chem., 1965,193,265.

3. Konisberg W. In Methods in Enzymology ; Hirs C.H.W.; Timasheff S.N. Eds.. Academic Press: New York. 1972, Vol. 25, p.185.

4. Lee M.S., Jen L.C., Clifford A.J., Whitaker J.R. and Feeney R.E. J. Agric. Food Chem., 1979, 27, 1094.

5. Habeeb A.F. Anal. Biochem., 1966, 14,328-336.

6. Hidalgo F.J. and Kinsella J.E.. J. Agric. Food Chem. 1989, 37, 860-866.

7. Spector A.A. and Fletcher J.E. Lipids, 1969, 5, 403.

8. Fulgate R.; Song P.S. Biochim. Biophys. Acta, 1980, 625, 28-42.

9. O'Neill T.E.; Kinsella J.E. J. Agric. food Chem. 1987, 35, 770-774.

10. Papiz M.Z.; Sawyer L.; EliopoulosE.E.; North A.C.T.; Findlay J.B.C.; Sivaprasadarao R.; Jones T.A.; Newcomer M.E.; Kraulis P.J.. Nature, 1986,324,383-385.

11. Dickinson E.;, Rolfe S.E.; Dalgleish D.G. Food Hydrocolloids. 1989, 3,193.

12. Closs B. Thèse de Doctorat de l'Universite de DIJON, DIJON, 1990.

13. Small D.S. In Handbook of Lipid Research- Plenum Press, New York, 1986, Vol. 4, p.345-394.

14. Cornell D.G.; Patterson D.L. J. Agric. Food Chem., 1989, 37, 1455-1459.

15. Colas B.;, Gobin C.; Lorient D.. J. Dairy Res., 1988, 55, 539-546.

RECEIVED May 16, 1990

Chapter 11

Some Aspects of Casein Micelle Structure

Lawrence K. Creamer

New Zealand Dairy Research Institute, Palmerston North, New Zealand

Casein, an important group of nutritional proteins in milk, exists in the form of colloidal particles, known as micelles, in the milks of all mammals examined so far. Several models for the structure of the micelle have been proposed. Much of the experimental data is explained by the models in which casein molecules form "submicelles" of calcium-sensitive caseins. These submicelles are held together with colloidal calcium phosphate and in which κ-casein resides largely on, or close to, the outer surface of the micelle. Any valid model of the casein micelle must satisfy a number of criteria including consistency with the data derived from examination of a number of micelle systems from several different mammalian species, the genetic coding information for the casein proteins, the recent spectral data, and the electron microscopic evidence.

All young mammals rely on milk for their sustenance and early growth. These milks contain lactose, fat, salts and proteins. These proteins are classified on their solubility at pH 4.6 as either whey proteins (soluble) or caseins (insoluble). The caseins generally exist in milk as small colloidal micellar particles and the structure of both the caseins and the micelles has been a matter of study for many years by a large number of groups. These micelle studies have been reviewed from time to time (1-6). In the first section a brief history of the development of one of the currently accepted models of the casein micelle is presented. This is followed by an outline of some of the criteria that a model should meet with comments on how the known data fulfill these criteria and then discussion on some unresolved problems.

0097–6156/91/0454–0148$06.00/0

Brief History

One of the early observations was that if skimmed milk was acidified a precipitate of casein formed and if this was dissolved at a pH of 7 a clear solution resulted. However on addition of calcium chloride solution, a micellar suspension was formed that had an opacity similar to that of skim milk. Waugh and his co-workers at MIT (Massachusetts Institute of Technology) found that seemingly similar micelles could be formed from the minimal system of α_{s1}- and κ-casein in a calcium-containing buffer solution. They then studied these micelles extensively and concluded, after a number of earlier proposals, that the casein micelle probably consisted of a core of calcium-insoluble casein surrounded by a coat, or monolayer, of κ-casein (or a 1:1 complex of κ-casein and α_s-casein) (*1, 2, 7-11*). (Note: α_s-Casein is a mixture of α_{s1}- and α_{s2}-caseins with their various levels of phosphorylation.) They continued to study both the "core" and "coat" structures (*9-11*) and incorporated the results and conclusions of other studies that had been carried out to conclude that the core was made up of submicelles and each of these contained a mixture of calcium-sensitive (now known as α_{s1}-, α_{s2}-, and β-) casein fractions (Table I). It was considered that these proteins commingled at the molecular level, a consideration that was omitted from the summary diagram (Fig. 1; from *9*). From the early studies by the MIT group using systems that contained no inorganic *ortho*-phosphate, the submicelles were considered to aggregate through calcium-mediated bonds. (Studies by others and (later) by Waugh and his colleagues (*2*) showed that inorganic phosphate played an important role in micelle formation and stability.)

Table I. Approximate Composition of Bovine Casein Micelles

Component	Concentration (mg/g)
Water	660
α_{s1}-Casein	111
α_{s2}-Casein	33
β-Casein	104
γ-Casein	8.7
κ-Casein	40
Calcium	12.7
ortho-Phosphate	10.6
Other components	20

Investigations of casein-casein interactions (*10-11*) suggested that the micelle coat consisted of a low ratio α_s- to κ-casein mixture. When Waugh reviewed the area in 1970 (2), he included new material relating to the aggregation behavior of submicelles in the presence of both calcium and phosphate.

In 1966 Payens published a possible model of the casein micelle (12) based on data relating to the known physical properties of the caseins and the casein micelle.

Morr published a model of the micelle in 1967 (13) which accounted for most of the known information at the time, including the behavior of casein micelles in milk after the action of the enzymes of the milk coagulant, rennet, and after milk had been heated when at least one of the whey proteins (β-lactoglobulin) condensed onto or into the casein micelle.

Slattery and Evard (14) re-examined the interactions of κ-casein with α_{s1}-casein and concluded that a better model for the casein micelle was one in which there was a range of submicelles each with different κ-casein contents and that these submicelles assembled into micelles so that the κ-casein-rich submicelles were on the micelle surface and κ-casein could be on the outside of the micelle.

Schmidt, on the basis of a study of the properties of micelles that had been assembled from micellar components in various ratios (15-17), proposed a model in which the submicelles were held together by calcium phosphate, possibly $Ca_9(PO_4)_6$, complexes (4, 18). It was suggested that the submicellar centers were largely hydrophobic, while the surfaces of the submicelles were hydrophilic and, in the absence of calcium, carried a high negative charge.

At about the same time a study on the effect of chymosin on the hydrodynamic diameters of casein micelles (19) supported the suggestion (20) that more emphasis should be given to the nature of the surface of the micelle and indicated that it could be considered as "hairy" with "hairs" of κ-casein macropeptide projecting into the solution and giving steric interactions with the surfaces of other micelles.

More recent studies have expanded on the "hairy" and the "coat-core with submicelles" models. The remainder of this paper will be devoted to high-lighting aspects of micelle structure in some detail.

Criteria for Model Selection

To be valid, any model of the casein micelle must fulfill certain criteria. Some of these are listed in Table II and will be treated as discussion points.

Nature of the Forces that Stabilize the Micelle. Implicit in the use of the name "micelle" for the small casein aggregates in milk is the suggestion that they are like other micelle systems, such as soaps or detergents, and are in a system with equilibrium between the micelles and the micellar components. There is considerable doubt about the equilibrium nature of the micelle as it is made up of more than one component and the amount of soluble casein depends on several factors, including protein concentration. Payens (12, 21) calculated that the measured charge on the casein micelle was insufficient to give stability to the micelle and concluded that one of the parameters in the calculation needed to

Table II. Some of the Criteria that a Micelle Model should Fulfill

---

1. Explain the high stability of the casein micelle to moderate changes in pH, salt concentration, temperature, and non-aqueous solvent concentration

2. Incorporate a reasonable explanation of the role of calcium and phosphate in micelle structure and stability

3. Explain how κ-casein is accessible for interaction with both rennet enzymes and denaturing whey proteins in heated milk

4. Be consistent with most of the electron microscopic evidence

5. Be consistent with the characteristics of the casein components and the relative strengths of their interactions with one another

6. Explain why stable casein micelles can be made with calcium or some other polyvalent cation and without colloidal phosphate

7. Explain the high, and pH dependent, water content of the casein micelle

8. Explain how micelles can be made with a relatively wide range of casein component concentrations

9. Explain why micelles from milks of different mammalian species seem to be similar despite their very different protein compositions

10. Explain why the rate of amino acid sequence divergence for the caseins can be so high without seeming to affect micelle structure

11. Be consistent with recent spectral data

---

be much smaller for the micelles to be stabilized by electrostatic repulsion. He decided that this was reasonable on the basis that the casein micelles were much more hydrated than the structures that the theory was designed to describe. Nevertheless there is some doubt that electrostatic repulsion is the main stabilizing force for the casein micelle.

κ-Casein seems to be of crucial importance in stabilizing the micelle and one of the important factors cited has been that chymosin cleavage of κ-casein gives two large peptides of very different properties. Para-κ-casein (105 amino acids) is insoluble, carries a net positive charge at pH 7, and strongly associates with the calcium-sensitive caseins while the "macropeptide" (64 amino acids) is soluble, even in moderate concentrations of trichloroacetic acid, carries a net negative charge and occupies a greater volume in buffer and in SDS solution than expected from its molecular weight. Addition of chymosin to milk causes the micelles to coagulate as the κ-casein is converted to para-κ-casein, essentially because of the changes to the micelle surface from macropeptide (hydrophilic with a negative charge) to para-κ-casein (hydrophobic with a positive charge). In the studies on casein micelle structure, the MIT group prepared "pseudo" micelles from α_s- and β-caseins in appropriate phosphate-free buffer solutions and then added κ-casein to these suspensions of colloidal aggregates (2, 11). The pseudo micelles were less stable to perturbations in calcium concentration, for example, than micelles. However they were more stable than micelles that had been made with κ-casein and then treated with chymosin.

On the basis of the decrease in volume of casein micelles as a consequence of chymosin action it has been suggested (19) that the micelle can be considered to be "hairy" with the principal stabilizing force being from the steric effects of the very hydrophilic layer on the surface of the micelle rather than the net negative charge of the micelle. The notion that κ-casein bound to the micelle has a portion of it's structure protruding from the micelle surface into the micelle solvent is supported by recent spectral evidence - see later.

In conclusion, there is no clear evidence that either the steric effect or the net negative charge on the micelle is the more important for stability. What is clear is that the presence of κ-casein is important.

Role of Colloidal Calcium Phosphate. Early studies (22, 23) demonstrated that if milk was acidified at low temperature and then dialyzed against cold milk for several days, the resultant material, called colloidal phosphate-free milk, differed from the original milk in a number of ways. Most importantly, it contained little colloidal calcium phosphate, was more viscous, was translucent in appearance and did not contain normal casein micelles, despite having the same concentration of calcium ions as normal milk (22, 23). A recent report (24) suggests that if the final stages of the dialysis are carried out at 30 °C (rather than 4 °C) then normal micelles were formed. This needs to be investigated further.

In studies using artificial systems (e.g. 2, 15-17) the addition of calcium and phosphate to mixtures of caseins show that they are incorporated into the micelle. The early model of Waugh (7, 8, 10) did not give calcium phosphate a specific role, but all more recent models (e.g. 2, 4, 18) include it in some way or another. At calcium ion concentrations, e.g. 10 mM, where casein in appropriate buffers formed micellar aggregates, micelles containing both calcium and phosphate were less hydrated (25).

In addition to dispersing casein micelles by removal of the calcium phosphate (26), they can be dispersed by addition of urea solution. The question arises as to what has happened to the colloidal calcium phosphate. One early study (27) showed that it did not pass through the dialysis sac during extensive dialysis at 4 °C while another (28) showed that 4 M urea solution did not disperse micelles into monomeric casein molecules in the presence of colloidal calcium phosphate.

A more recent study (*29*) separated the colloidal material and its associated protein in 6 M urea solution using HPLC techniques.

Numerous studies have been undertaken into the reactions of calcium, phosphate, citrate, etc. in solution and the complexes that form. The most recent suggestions (*30*) are that micellar calcium phosphate is a complex of the phosphoserine clusters of the caseins with an acidic amorphous calcium phosphate. The most appropriate simple model compound is brushite, $CaHPO_4.2H_2O$. It is possible that the casein chains near the phosphorylation sites are α-helical in structure (*31*).

The necessity for a structural role for colloidal calcium phosphate seems unclear, because micelles of some sort can form, albeit at higher calcium ion concentrations in the absence of inorganic phosphate. Perhaps, with milk having its nutritional role, one function for the casein is to be a carrier of calcium and phosphate for adequate skeletal mineralization for the neonate.

Accessibility of κ-Casein. When casein micelles are reacted with milk-coagulating enzymes, κ-casein is converted to para-κ-casein by enzymatic cleavage of a single peptide bond. The micelles then aggregate to form a clot (coagulum). The κ-casein must be accessible to the enzyme for this reaction to occur.

When milk is heated, some of the whey proteins denature and form disulfide bonded complexes with κ-casein (*32*). Again the κ-casein has to be accessible to the denaturing proteins at the temperature of denaturation. All of the models mentioned in this review put κ-casein on the external surface of the micelle. There are a number of other proposals in which the κ-casein is distributed more evenly throughout the micelle. In these cases the proposers suggested that rennet enzymes could travel through a hydrated and relatively porous micelle. The problem of how whey proteins in milk heated to temperatures above their denaturation point could associate with κ-casein was not adequately addressed.

A proportion of the κ-casein in bovine milk and micellar systems is in the whey/supernatant and seems to be readily dissociated from the micelle surface at low temperatures but may not be in equilibrium with that on the micelle (*33*). Clearly such protein would also react with enzymes or partly denatured whey proteins and may well then associate with the micelle surface because the para-κ-casein or the complex would have different interaction characteristics from those of κ-casein. Incidentally assays for κ-casein in the bovine system are relatively difficult because of the post-translational heterogeneity of the protein.

A further complication is that bovine κ-casein contains two cysteine residues and as isolated from the natural micelle system κ-casein is disulfide crosslinked into polymers of different sizes including some very large polymers (*34*). The implications of these findings are that each of these large polymers covers a moderate proportion of the micelle surface, possibly spanning several submicelles or that even mild isolation procedures induce disulfide linkage. Explorations to try to determine why the bovine κ-casein is polymeric in the native micelle should be a useful area of study.

Electron Microscopic Data. The model-builders have made extensive use of the results from electron microscopic studies. The earliest of these showed that micelles were small, heterogeneously sized, essentially spherical protein particles (*35, 36*). More sophisticated techniques showed that there was likely to be a sub-structure to the micelle and that there were distinct regions of electron-dense material, probably rich in calcium and phosphate (*37*). Freeze-fracture techniques showed quite clearly that there was an internal structure to the micelle and supported the earlier electron microscopic work (*38*). Two recent rotary shadowing studies; one using intact micelles (*39*) the other with freeze-fractured micelles (*40*), showed that the submicelles are heterogeneous in size and may be far from spherical in shape. Examination of the internal structure of the submicelle has been interpreted to show (*41*) that the casein molecules seem to take the shape of 1-3 nm diameter coils that are wound up within the submicelle. Electron microscopic studies of heated milks indicate that the denatured whey protein/κ-casein complex is on the surface of the micelle under appropriate pH conditions (*39, 42*). However, electron microscopy does not seem to have been able to place κ-casein on the micelle surface with certainty, despite several attempts.

For the micelle models to be consistent with the above data, they need to incorporate a submicelle structure and to localize the mineral material.

Casein Characteristics. A number of studies were carried out to characterize the caseins shortly after they were obtained in pure form. Other studies were carried out on the ways in which the casein components interacted with one another. Whole bovine casein comprises four major caseins; namely, α_{s1}-, α_{s2}-, β-, and κ-casein and each has different properties (*43*).

α_{s1}-**Casein.** This protein consists of 199 amino acids, 8 of which are phosphoserine. It shows strong self-association behavior and is readily precipitated at low (4-5) pH or with calcium salts at neutral pH.

α_{s2}-**Casein.** This protein consists of 207 amino acids with between 10 and 13 phosphoserines. It shows a strong self-association and will precipitate at lower calcium chloride concentrations than α_{s1}-casein.

β-**Casein.** This protein consists of 209 amino acids, 5 of which are phosphoserine. It is readily precipitated at low pH or with calcium salts at neutral pH and at higher (>8 °C) temperatures. Its self-association is of the micellar type where there co-exists both monomers and large (about 40-mer) polymers, but few dimers, trimers, etc. Its self-association is also temperature dependent so that at about 5 °C, neutral pH, moderate ionic strength (ca. 0.08), and moderate concentration (ca. 1mg/mL) it is not precipitated with calcium although it is reversibly precipitated under these solution conditions at higher temperatures. Under physiological conditions α_{s1}-casein associates more strongly with β-casein than either protein self-associates.

κ-**Casein.** This protein consists of 169 amino acids and the chain has a different character at the two ends; the amino end is more hydrophobic and carries more positive charges than the carboxyl end. This amphiphilic character is evident in β-casein as well, but is more pronounced in κ-casein than almost any other protein. Pure κ-casein does not precipitate at low pH or in the presence of calcium salts between 0 and 50 °C and between pH 2 and 10. However it associates strongly with the other caseins and in admixture with them it is readily precipitated either at low pH or with calcium salts at neutral pH. In a mixture of α_{s1}- and κ-casein, α_{s1}-:κ-casein complexes form in preference to either α_{s1}-:α_{s1}- or κ-:κ- casein complexes. As mentioned above α_{s1}-:β- complexes are stronger than α_{s1}-:α_{s1}- complexes. The relative strength of these associations are not as well known in systems that contain colloidal calcium phosphate, but it is possible that the interactions among the calcium-sensitive caseins are stronger in the presence of colloidal calcium phosphate than those between κ-casein and the other caseins.

Ono et al. (*44*) dissociated bovine casein micelles by removing colloidal calcium phosphate and examined the casein complexes using HPLC and found that they could distinguish a number of specific complexes. The major earlier eluting fraction (F2) contained a 1:1 ratio of α_s- to κ-casein while the other major fraction (F3) contained a 1:1 ratio of α_{s1}- to β-casein. Under the conditions of analysis (25 °C, pH 7.0, 0.07 M NaCl) these complexes were quite stable. This result verifies the conclusions derived from the studies on mixtures of individual components (*2, 45*).

Micelles with Divalent Cations and without Colloidal Phosphate. The accepted structure of the casein micelle gives colloidal calcium phosphate an important role in cementing the casein submicelles together (see above). Nevertheless casein micelles can be made in the minimal system of α_{s1}- and κ-casein with calcium chloride, sodium chloride and a neutral pH buffer (*2*). In these systems the calcium ion concentration is higher under conditions where micelles form than in normal bovine milk. Examination of the binding of calcium to the casein components in the presence of phosphate (Waugh and Creamer, unpublished, 1965) indicated that phosphate affected calcium binding only after colloidal casein particles were formed. A decrease in the binding of water by the phosphate-containing system was also noted. Similar results were obtained with whole casein (*25*). These results suggested that artificial systems could exhibit the same properties as those shown by colloidal phosphate-free milk.

Clearly the micelles in the phosphate-free system are different from natural micelles and the details of their structure may be quite different. Nevertheless the calcium-sensitive caseins aggregate and coalesce into submicelles and these associate under the influence of calcium and κ-casein aggregates onto, or becomes incorporated into, the surface of the micelle. Presumably a similar set of reactions occurs when whole casein is mixed with other ions such as barium or manganese.

Micelle Solvation. Waugh noted that casein micelles contained large amounts of water (*7*) and that it was dependent on the level of calcium bound by the casein (*9, 11, 46*). As mentioned above, phosphate in the micelle decreases the water content. Micellar water content is dependent on milk pH and has been measured and correlated with heat stability in a series of milks from individual cows (47). (This conclusion has not been verified in later studies (48, 49).)

There are three types of water in the casein micelle: the water bound to each protein that constitutes the micelle by the ionic, etc. groups on the protein; the water that is in the interstices between the protein molecules of the submicelles, and the water that is between the submicelles within the micelle. Although there are no current theories on why the casein micelles are so hydrated, it is likely that hydration is governed by the way in which the casein molecules fold into submicelles, which, in turn is governed by the primary sequences of the calcium-sensitive caseins. If a very simplistic model is assumed, i.e. that the micelle is an open spongy structure with few crosslinks and with charged groups spread throughout; then the major factors are explainable on an electrostatic basis. For example the effect of pH on hydration (increases away from the iso-electric point) or the effect of calcium binding to reduce hydration. Ion-exchangers with a low degree of crosslinking behave in a comparable fashion. It seems more likely, however, that a major contribution is made by the way in which the protein strands fold to maintain an open structure, and that these electrostatic contributions are rather less important.

Micelles with Different Ratios of Casein Components. Early studies on natural bovine micelles showed that the quantity of each casein component in bovine milk was almost invariant from animal to animal in mid-lactation and during the major part of the lactation (50). (Since that time it has been found that there are variations in the quantity of κ-casein in the milk of any single cow and that it is related to a specific genetic variant (51).) A further early finding was that, within any one milk sample, micelle size and κ-casein content were inversely correlated (52). This was also true for the minimal casein micelles made with α_{s1}- and κ-casein with calcium chloride in a neutral buffer solution (2). Schmidt and his colleagues found that they could make casein micelles with a wide range of compositions that were stable under a number of different conditions (15-17). (They were also able to support earlier contentions that micelles made from mixtures containing more κ-casein were smaller.) This suggested that the structure of the micelle was not particularly sensitive to its protein composition and suggests that submicelles can be made with almost any ratio of different calcium-sensitive caseins without any great change in micelle properties. When the calcium-sensitive caseins come together to form the core of the micelle the quantity of colloidal calcium phosphate may be governed by the level of phosphorylation of the caseins. In a recent study (53) where the phosphorylation level was altered by enzymic and chemical means, the micelles with fewer phosphorylated residues contained, somewhat surprisingly, more colloidal calcium phosphate and were less stable, but had the same sub-structure as seen by freeze-fracture electron microscopy. Clearly more work is required in this area.

Micelles in Milk of Other Species. Examination of the casein compositions of the milks of various species of mammals showed that the ratios of α_s-type to β-type caseins was quite different. Table III shows examples of the casein composition of several milks. Also shown in parenthesis is the ratio of the quantity of each casein to β-casein. Clearly the human milk contains very little α_s-casein (*54-56*), and yet all of the casein systems form micelles with rather similar structures as seen by electron microscopy (*56, 57*). The goat is especially interesting because it is closely related to both the sheep and the cow and yet in some goat milk there is virtually no protein that corresponds to bovine α_{s1}-casein (*58*).

Table III. Approximate Concentration (g/L) of Casein Components in Various Milk Types

Casein	Milk Type			
	Human	Cow	Sheep	Goat
α_s-[a]	<0.3 (0.17)[b]	14 (1.27)	25 (1.0)	2 or 9 (0.11 or 0.50)[c]
β-	1.8 (1.00)	11 (1.00)	25 (1.00)	18 (1.00)
κ-	0.6 (0.33)	4 (0.36)	10 (0.40)	4 (0.22)

[a]Mixture of α_{s1}- and α_{s2}-caseins at all levels of phosphorylation.
[b]Values in parenthesis are ratios to β-casein.
[c]The quantity of α_{s1}-casein in goat casein is genetically controlled and can vary from nil to ca. 3.6 g/L for each variant (*63*). The quantity of α_{s2}-casein is not as variable and is ca. 2 g/L.

It might have been supposed, and this was the premise for studies carried out in our laboratory (*57-62*) and by others, that if the structure of the micelle was important in a biological sense, then if there were changes in the ratio of the casein components without changes in the structure of the casein micelles, it would be expected that there should be compensating changes in the composition and molecular characteristics of the proteins. Our studies showed that there were few differences between the micelles from the sheep, goat and cow milks despite the apparent absence of an α_{s1}-casein from the goat milk we investigated (*62*). Although the sequences of the corresponding proteins from each species are different (see next section) there did not seem to be any compensating characteristics, i.e. the β-caseins of goat (or human) milk behave more like bovine β-casein than an α_s-casein. As discussed in the previous section, these data suggest that, within certain limits, the structure of the micelle is not very sensitive to the ratio and composition of the calcium-sensitive caseins that are incorporated into its structure.

It is noteworthy that all mammalian milks studied so far have a κ-casein and a β-casein and that the κ-casein is about 15% of the total casein. It will be interesting to compare the data available on mammalian milks with that from marsupial (primitive mammal) milks. The other notable factor is that all milks seem to contain colloidal calcium phosphate (*54*). Examination of micelle formation with human caseins (*64*) indicated that the β-caseins with greater numbers of phosphate residues (3 or 5) gave micelles under the conditions of study whereas the β-casein with fewer phosphate residues did not. These data suggest that κ-casein and β-casein (with five phosphates) is the minimal protein system for micelle development. A more speculative suggestion is that in systems that contain β-casein, this protein is important in the formation of submicelles and that the α_s-caseins are of lesser importance and become integrated into the β-casein detergent-type submicellar structure.

Rate of Divergence of the Casein Sequences. An early study showed that the sequences of the C-terminal portions (macropeptides) of the κ-caseins of a number of different species were quite variable (*65*). More recent studies (*66-68*) have found that the genes coding for the caseins have a high mutation rate in all the coding region except that which codes for the residues near the phosphorylation sites in α_s- and β-caseins and the chymosin site in κ-caseins. This finding fits with the notion that the phosphoserine residues of the calcium-sensitive caseins are involved in the aggregation of the submicelles into micelles by way of colloidal aggregates of calcium phosphate and that this is an important function. All milks appear to have a casein with properties similar to that of bovine κ-casein, i.e. very sensitive to chymosin/pepsin enzymes, soluble in calcium chloride solution and able to form strong complexes with calcium-sensitive caseins. Despite the findings that the casein mutation rate is very high it seems likely that there must be some characteristics of the caseins that are important for the natural function of digestibility. Micelle formation with stable interactions between the calcium-sensitive caseins and κ-casein are apparently preserved through the mutational changes. On the basis of the earlier discussion, it seems likely that there are some other regions where there can be some changes in the residues as long as the region of peptide can still function to maintain an open and non-globular structure. Whether there need to be regions that are moderately or strongly hydrophobic is not clear. The A variant of bovine α_{s1}-casein does not contain a sequence of 13 amino acids that give a hydrophobic region to the molecule, but this does not affect micelle formation, stability, etc. sufficiently for this genetically different casein to be selected against. On the same theme, it was found that bovine β-casein self-association (*69*) and association with α_{s1}-casein (*70*) was strongly affected by a short C-terminal sequence of hydrophobic residues. No doubt with the advent of genetic manipulations it will be possible to produce caseins devoid of such regions and observe the effect of these changes on the consequent milks.

Spectral Studies. Earlier studies (e.g. *71-73*) used optical methods (UV, IR, ORD, and CD) while more recent studies (*74-80*) on the caseins and casein micelles have concentrated on NMR and X-ray methods. The CD spectral data were used in two studies (*81, 82*) as the basis for the application of various predictive algorithms for estimating where various types of regular repeating structure might occur in the protein sequences. These algorithms have recently been used to examine caseins from several species in considerable detail (*31*).

One very important criticism of this type of analysis is that the original data used to derive the prediction parameters came from globular proteins, and it seems likely that the folding of the caseins, like the folding of most structural proteins, is on a different basis from that in the globular proteins where the prime prerequisite is for a stable skeleton whereon and wherein to place the various groups that are important for the particular function of that protein. This is almost the opposite from the casein structures which should be enzyme permeable, readily digested, and definitely not stable in the long-term.

Small angle neutron scattering studies (74) showed that casein submicelles, obtained by dialyzing a micelle fraction from milk against dilute NaCl solution, and micelles both gave scattering curves that were consistent with the dispersions containing spherical particles with an average diameter of 16.8 nm. Further studies (77) showed that the neutron scattering data was the result of protein structural order rather than colloidal calcium phosphate order. There was also some evidence for polydispersity in the size distribution of the submicelles.

Small angle X-ray scattering studies (78) made on sodium caseinate aggregates (submicelles) and calcium caseinate micelles that had been pelleted could be interpreted to show that the micelles were made up from submicelles (diameter about 20 nm) and that each submicelle had a core (diameter about 8 nm) of less hydrated material and an outer region of more hydrated material. The absolute values of hydration in that particular study seemed to be higher than those derived by other experimental methods and it seems likely that the absence of colloidal calcium phosphate from the system could have been a factor.

NMR data showed (75, 79, 80) that only some of the residues of the caseins were exposed to the solvent so that they were rotating freely. It appeared that the C-terminal one-third of κ-casein was in such a state both in κ-casein solution and in the casein micelle. Very few of the residues of the other casein components were in a freely rotating state at moderate temperatures.

Incompletely Resolved Problems

While most researchers consider the basic elements of the structure of the casein micelle, e.g. the importance of colloidal calcium phosphate, are well-founded there remain areas where further work is needed to resolve some uncertainties, for example: "Is κ-casein incorporated into the micelle or is it on/close to the micelle surface ?", and "Are there specific interactions between the various sidechains of the caseins that are important in the stabilization of micelle structure ?"

Siting of κ-Casein. Waugh and his co-workers (2) concluded that the casein micelle was made up of mixed α_s and β-casein subunits and that as these condensed with calcium (in model systems) or colloidal calcium phosphate, κ-casein (or 1:1 ratio α_s-:κ-casein mixtures) adsorbed onto the outer surface of these aggregates. The difficulty with this model is that it is known that α_s- and κ-casein form a strong complex so how can the κ-casein be available for adsorption onto the final micelle but not get caught up into the condensing micelle as the κ-:α_s-casein complex ? Perhaps, and Waugh recognized this, κ-casein was bound to α_s-casein partly through electrostatic bonding between the phosphoserine residues of the α_s-casein and the basic residues of κ-casein. In this case maybe the κ-casein is bonded to the α_s-casein until the α_s-casein becomes bonded into the colloidal phosphate complexes of the micelle, in other words perhaps κ-casein was displaced by calcium.

An alternative model, such as that proposed by Slattery and Evard (*14*), in which the submicelles are stable entities that contain a variable number of κ-casein molecules; some with none and some with many, requires that the κ-casein molecules are either on one side of the complex or at least they are free to move to one side of the complex as the submicelle is bound into the micelle. It appears that this model is not compatible with a high proportion of the κ-casein being polymeric with stable disulfide bonds.

A third possibility is that κ-casein is very mobile and in milk serum (or casein solution) it is present in submicelles, is present as κ-casein polymers, and is present as monomers.

Specific Sidechain Interactions. The overall structure of most globular proteins is such that there is a core of hydrophobic residues with a surface region of more hydrophilic residues. The amino acid residues in such a protein may be in one of a number of identifiable   structures and both the sidechain-sidechain interactions (ionic and hydrophobic bonding) as well as mainchain interactions (hydrogen bonding) contribute to the stability of these structures. The water surrounding the protein is an important factor in determining overall structure.

On the premise that protein structure is a reflection of function, it should be expected that caseins would not be similar to typical globular proteins. Whereas a globular protein often has some specific function with rigid spatial requirements, the main function of the caseins is to be readily digested to peptides and amino acids. Thus it has different structural requirements and does not need to have a rigid internal structure to perform its function. The question of whether casein needs any structural motifs other than the phosphate clusters for it to form into the observed submicelles and micelles is not easy to answer on the basis of the information presented above. On one hand the genetic data suggest that there is very little requirement for the protein to conserve its sequence of amino acids. (An extreme example is that of a major rat casein which contains nearly 60 extra amino acids as a consequence of the insertion of 9 or 10 near repeats of a hexapeptide.) On the other hand, the structure of the casein micelle, the ready digestibility of the casein, its acid precipitation, etc. all suggest that there must be some structural features that are important in a Darwinian selection sense. This suggests that there must be some underlying structural motifs that are larger or at a higher level than conservation of the nucleotides or even of the type (e.g. hydrophobic) of amino acid residue so that there can be such diversity at the molecular level.

κ-Casein, whether it is on the outside or integrated into the outer layers of the submicelles, must interact with the calcium-sensitive caseins. There is some data to support the notion that the positively charged region of κ-casein might react with the strongly negatively charged region of the calcium-sensitive caseins. However there may be more selective reactions involving some specific sites. Clearly more work needs to be carried out at a molecular level.

Conclusion

Most of the evidence points to the micelles from milks from different mammals being rather similar despite differences in casein type and composition. All of the micelles contain some colloidal calcium phosphate and it has an important role in stabilizing the structure of the native micelle. Only κ- and β-casein are present in all milks and the ratio of κ- to calcium-sensitive caseins seems to be similar in the milks examined so far. All micelles have κ-casein on (or in) the exterior surface, with the calcium-sensitive caseins in the interior and assembled into submicelles with an average diameter of between 10 and 20 nm. The κ-casein is essential for stabilizing the micelle against changes in its environment. The application of spectroscopic methods has verified and refined the earlier conclusions. Recent research in the molecular biology area points to site-directed mutagenesis and/or transgenic animals being techniques that have the potential to allow substantial progress to be made.

Literature Cited

1. Rose, D. *Dairy Sci. Abstr.* **1969**, *51*, 171-175.
2. Waugh, D. F. In *Milk Proteins*; McKenzie, H. A., Ed.; Academic Press: New York, 1971; Vol 2, pp 3-85.
3. Slattery, C. W. *J. Dairy Sci.* **1976**, *59*, 1547-1556.
4. Schmidt, D. G. In *Developments in Dairy Chemistry*; Fox, P. F., Ed; Applied Science Publishers, Essex, 1982; Vol. 1, pp 61-82.
5. McMahon, D. J.; Brown, R. J. *J. Dairy Sci.* **1984**, *67*, 499-512.
6. Farrell, H. M. Jr. In *Fundamentals of Dairy Chemistry*; Wong. N. P. Ed.; Van Nostrand Reinhold Company: New York, 1988; pp 461-510.
7. Waugh, D. F.; Noble, R. C. *J. Am. Chem. Soc.* **1965**, *87*, 2246-2257.
8. Noble, R. C.; Waugh, D. F. *J. Am. Chem. Soc.* **1965**, *87*, 2236-2245.
9. Waugh, D. F.; Creamer, L. K.; Slattery, C. W.; Dresdner, G. W. *Biochemistry* **1970**, *9*, 786-795.
10. Talbot, B.; Waugh, D. F. *Biochemistry* **1970**, *9*, 2807-2813.
11. Waugh, D. F.; Talbot, B. *Biochemistry* **1971**, *10*, 4153-4162.
12. Payens, T. A. J. *J. Dairy Sci.* **1966**, *49*, 1317-1324.
13. Morr, C. V. *J. Dairy Sci.* **1967**, *50*, 1744-1751.
14. Slattery, C. W.; Evard, R. *Biochim. Biophys. Acta* **1973**, *317*, 529-538.
15. Schmidt, D. G.; Koops, J.; Westerbeek, D. *Neth. Milk Dairy J.* **1977**, *31*, 328-341.
16. Schmidt, D. G.; Koops, J. *Neth. Milk Dairy J.* **1977**, *31*, 342-351.
17. Schmidt, D. G.; Both, P.; Koops, J. *Neth. Milk Dairy J.* **1979**, *33*, 40-48.
18. Schmidt, D. G. *J. Dairy Res.* **1979**, *46*, 351-355.

19. Walstra, P.; Bloomfield, V.A.; Wei, G. J.; Jenness, R. *Biochim. Biophys. Acta,* **1981**, *669*, 258-259.
20. Walstra, P. *J. Dairy Res.* **1979**, *46*, 317-323.
21. Payens, T. A. J. *J. Dairy Res.* **1979**, *46*, 291-306.
22. Pyne, G. T.; McGann, T. C. A. *J. Dairy Res.* **1960**, *27*, 9-17.
23. Pyne, G. T. *J. Dairy Res.* **1962**, *29*, 101-130.
24. Ono, T.; Obata, T. *J. Dairy Res.* **1989**, *56*, 453-461.
25. Creamer, L. K.; Yamashita, Y. *N. Z. J. Dairy Sci. Technol.* **1977**, *11*, 257-262.
26. Lin, S. H. C.; Leong, S. L.; Dewan, R. K.; Bloomfield, V. A.; Morr, C.V. *Biochemistry,* **1972**, *11*, 1818-1821.
27. McGann T. C. A.; Fox, P. F. *J. Dairy Res.* **1974**, *41*, 45-53.
28. Morr, C. V. *J. Dairy Sci.* **1967**, *50*, 1744-1751.
29. Aoki, T.; Kako, Y.; Imamura, T. *J. Dairy Res.* **1986**, *53*, 53-59.
30. Holt, C.; Kemenade, M. J. J. M. van; Nelson, L. S. Jr.; Sawyer, L.; Harries, J. E.; Bailey, R. T.; Hukins, D. W. L. *J. Dairy Res.* **1989**, *56*, 411-416.
31. Holt, C.; Sawyer, L. *Protein Engineering* **1988**, *2*, 251-259.
32. Singh, H. *N. Z. J. Dairy Sci. Technol.* **1988**, *23*, 257-273.
33. Rose, D. *J. Dairy Sci.* **1968**, *51*, 1897-1902.
34. Parry, R. M.; Carroll, R. J. *Biochim. Biophys. Acta* **1969**, *194*, 138-150.
35. Nitschmann, H. *Helv. Chim. Acta* **1949**, *32*, 1258-1264.
36. Hostettler, H.; Imhof, K. *Milchwissenschaft* **1951**, *6*, 351-354, 400-402.
37. Shimmin, P. D.; Hill, R. D. *Aust. J. Dairy Technol.* **1965**, *20*, 119-122.
38. Schmidt, D. G.; Buchheim, W. *Milchwissenschaft* **1970** 25, 596-600.
39. Harwalkar, V.R.; Allan-Wojtas, P.; Kalab, M. *Food Microstructure* **1989**, *8*, 217-224.
40. Heertje, I.; Visser, J.; Smits, P. *Food Microstructure* **1985**, *4*, 267-277.
41. Kimura, T.; Taneya, S.; Kanaya, K. *Milchwissenschaft* **1979**, *34*, 521-524.
42. Creamer, L. K.; Berry, G. P.; Matheson, A. R. *N. Z. J. Dairy Sci. Technol.* **1978**, *13*, 9-15.
43. Swaisgood, H. E. In *Developments in Dairy Chemistry*; Fox, P. F., Ed; Applied Science Publishers, Essex, 1982; Vol. 1, pp 1-59.
44. Ono, T.; Odagiri, S.; Takagi, T. *J. Dairy Res.* **1983**, *50*, 37-44.
45. Creamer, L. K.; Berry, G. P. *J. Dairy Res.* **1975**, *42*, 169-183.
46. Waugh, D. F.; Slattery, C. W.; Creamer, L. K. *Biochemistry*, **1971**, *10*, 817-823.
47. Thompson, M. P.; Boswell, R. T.; Martin, V.; Jenness, R.; Kiddy, C. A. *J. Dairy Sci.* **1969**, *52*, 796-798.
48. Sood, S. M.; Sidhu, K. S.; Dewan, R. K. *N. Z. J. Dairy Sci. Technol.* **1979**, *14*, 217-225.
49. Morrissey, P. A.; Murphy, M. F.; Hearn, C. M.; Fox, P. F. *Irish J. Food Sci. Technol.* **1981**, *5*, 117-127.
50. Davies, D. T.; Law, A. J. R. *J. Dairy Res.* **1980**, *47*, 83-90.
51. Sullivan, R. A.; Fitzpatrick, M. M.; Stanton, E. K. *Nature* **1959**, *183*, 616-617.
52. McLean, D. M.; Graham, E. R. B.; Ponzoni, R. W.; McKenzie, H. A. *J. Dairy Res.* **1984**, *51*, 531-546.
53. Schmidt, D. G.; Poll, J. K. *J. Neth. Milk Dairy J.* **1989**, *43*, 53-62.

54. Jenness, R. In *Developments in Dairy Chemistry*; Fox, P. F., Ed; Applied Science Publishers, Essex, 1982; Vol. 1, pp 87-114.
55. Kunz, C.; Lonnerdal, B. *In Protein and Non-Protein Nitrogen in Human Milk*; Atkinson, S. A.; Lonnerdal, B., Eds; CRC Press Inc.: Boca Raton, FL,1989; pp 3-27.
56. Ruegg, M.; Blanc, B. *Food Microstructure* 1982, *1*, 25-47.
57. Richardson, B. C.; Creamer, L. K.; Pearce, K. N.; Munford, R. E. *J. Dairy Res.* 1974, *41*, 239-247.
58. Richardson, B. C.; Creamer, L. K. *Biochim. Biophys. Acta* 1975, *393*, 37-47.
59. Richardson, B. C.; Creamer, L. K.; Munford, R. E. *Biochim. Biophys. Acta* 1973, *310*, 111-117.
60. Richardson, B. C.; Creamer, L. K. *Biochim. Biophys. Acta* 1974, *365*, 133-137.
61. Richardson, B. C.; Creamer, L. K. *N. Z. J. Dairy Sci. Technol.* 1976, *11*, 46-53.
62. Ono, T.; Creamer, L. K. *N. Z. J. Dairy Sci. Technol.* 1986, *21*, 57-64.
63. Grosclaude, F.; Mahe, M. F.; Brignon, G.; Stasio, L. Di; Jeunet, R. *Genetique, Selection, Evolution* 1987, *19*, 399-411; *Dairy Sci. Abstr.* 1988, *50*, abstr 5258.
64. Azuma, N.; Kaminogawa, S.; Yamauchi, K. *Agric. Biol. Chem.* 1985, *49*, 2655-2660.
65. Mercier, J-C.; Chobert, J-M.; Addeo, F. *FEBS Letters* 1976, *72*, 208-214.
66. Stewart, A. F.; Willis, I. M.; MacKinlay, A. G. *Nucleic Acids Res.* 1984, *12*, 3895-3907.
67. Yu-Lee, L-Y.; Richter-Mann, L.; Couch, C. H.; Stewart, A. F.; MacKinlay, A. G.; Rosen, J. M. *Nucleic Acids Res.* 1986, *14*, 1883-1902.
68. Bonsing, J.; MacKinlay, A. G. *J. Dairy Res.* 1987, *54*, 447-461.
69. Thompson, M. P.; Kalan, E. B.; Greenberg, R. *J. Dairy Sci.* 1967, *50*, 767-769.
70. Berry, G. P.; Creamer, L. K. *Biochemistry* 1975, *14*, 3542-3545.
71. Herskovits, T. T. *Biochemistry* 1966, *5*, 1018-1026.
72. Ho, C.; Waugh, D. F. *J. Am. Chem. Soc.* 1965, *87*, 889-892.
73. Creamer, L. K. *Biochim. Biophys. Acta*, 1972, *271*, 252-261.
74. Stothart, P. H.; Cebula, D. J. *J. Mol. Biol.* 1982, *160*, 391-395.
75. Griffin, M. C. A.; Roberts, G. C. K. *Biochem. J.* 1985, *228*, 273-276.
76. Griffin, M. C. A. *J. Colloid Interface Sci.* 1987, *115*, 499-506.
77. Stothart, P. H. *J. Mol. Biol.* 1989, *208*, 635-638.
78. Pessen, H.; Kumosinski, T. F.; Farrell, H. M. Jr. *J. Dairy Res.* 1989, *56*, 443-451.
79. Rollema, H. S.; Brinkhuis, J. A.; Vreeman, H. J. *Neth. Milk Dairy J.* 1988, *42*, 233-248.
80. Rollema, H. S.; Brinkhuis, J. A. *J. Dairy Res.* 1989, *56*, 417-425.
81. Loucheux-LeFebvre, M-H.; Aubert, J-P.; Jolles, P. *Biophys. J.* 1978, *23*, 323-336.
82. Creamer, L. K.; Richardson, T.; Parry, D. A. D. *Arch. Biochem. Biophys.* 1981, *211*, 689-696.

RECEIVED October 15, 1990

Chapter 12

Cross-Linkage Between Casein and Colloidal Calcium Phosphate in Bovine Casein Micelles

Takayoshi Aoki

Department of Animal Science, Faculty of Agriculture, Kagoshima University, Kagoshima 890, Japan

The existence of cross-linkage between casein and colloidal calcium phosphate (CCP) has been proved by separating casein aggregates cross-linked by CCP from bovine casein micelles. Not only calcium and phosphate but also citrate play an important role in cross-linking of casein by CCP. The incorporation rates of individual casein constituents into casein aggregates cross-linked by CCP were in the order α_{s2}- > α_{s1}- > β-casein. This order was the reverse of the order of the dissociation rates of casein aggregates cross-linked by CCP during dialysis. The higher the ester phosphate content, the faster was the incorporation rate of the individual casein constituent and the slower is the dissociation rate. Ester phosphate groups were identified as the sites for interaction with CCP since α_{s1}-, α_{s2}- and β-casein were cross-linked by CCP while κ- and γ-casein and dephosphorylated α_{s1}-casein were not cross-linked. Relation between CCP cross-linkage and changes in casein micelles on heating and cooling was also described.

Bovine casein micelles are spherical colloidal particles of 20-600 nm in diameter, which are heterogeneous in constitution (1). They are composed of 93% casein and 7% inorganic constituents. The major casein constituents are α_{s1}-, α_{s2}-, β- and κ-casein, in the proportions of 3 : 0.8 : 3 : 1. The main inorganic constituent of casein micelles is colloidal calcium phosphate (CCP).

In bovine milk, about two-thirds of the calcium and half of the inorganic phosphate are present in the colloidal form with remainder in the soluble form (2). Skim milk is considered a two-phase system of calcium phosphate in quasi-equilibrium with an aqueous solution of salts and proteins (3). It has been

0097–6156/91/0454–0164$06.00/0

recongnizedthat CCP is an important constituent in maintaining the integrity of casein micelles since casein micelles are disaggregated into submicelles when CCP is removed (4).

Schmidt (1) expanded Slattery's model (5) for the structure of casein micelles and proposed a new model in which casein micelles are composed of submicelles held together by means of CCP. In this model, submicelles are heterogeneous in their composition, especially in κ-casein content. Submicelles rich in κ-casein are at the surface of micelles and submicelles poor in κ-casein are in the interior side of micelles. They are linked together by CCP which is represented as the formula of $Ca_9(PO_4)_6$. Although the sites of casein for interaction with CCP was assumed to be phosphate groups or carboxyl groups (6), they were not identified. There was no experimental evidence for the linkage between casein and CCP.

It is recognized that CCP participates in the changes of casein micelles in milk processing such as heating, cooling and rennet coagulation. Heating milk causes an increase in CCP level in bovine milk (2,3) and it is suggested that CCP is responsible for variations in the heat stability of milk (7). Cooling milk causes a decrease in the CCP level and the release of serum casein from casein micelles (8-11). Therefore, it is very important to elucidate the role of CCP.

Separation and Properties of Casein Aggregates Cross-linked by CCP

Casein micelles are disaggregated without dissolution of CCP when urea is added to milk to give a high concentration (12). However, when casein micelles separated by ultracentrifugation are solubilized in a concentrated urea solution, CCP is considered to be partially solubilized. Accordingly, in the present study 6 M urea simulated milk ultrafiltrate (USMUF) was used to disaggregate casein micelles (13). The casein micelles obtained by centrifugation of bovine skim milk at 100000×g for 1h at 25 °C were dissolved in USMUF at a 2.4% casein concentration. When high performance gel chromatography of bovine casein micelles disaggregated in USMUF was carried out using a TSK-GEL G4000SW column, the eluate was divided into two fractions, 1 and 2 (Figure 1a). The fraction eluted after fraction 2 was assigned to non-casein components contained in sedimented casein micelles. In order to identify each casein fraction, gel chromatography of individual casein constituents was carried out. α_{s1}-, α_{s2}- and β-Casein and reduced κ-casein were eluted as a single peak with a similar retention time to that of the main peak of fraction 2. When casein micelles were reduced with 10 mM 2-mercaptoethanol, fraction 1 decreased and fraction 2 increased (Figure 1b). The relative percentages of fractions 1 and 2, which were determined from the peak area of the chromatogram, were 57.3% and 42.7%, respectively. 2-Mercaptoethanol eluted just after non-casein components. Fraction 1 was not observed in reduced colloidal phosphate-free micelles (Figure 1c). The elution pattern of Figure 1c was the same as that of whole acid casein. Accordingly, it is suggested that fraction 1 of reduced casein micelles consists of casein aggregates cross-linked by CCP.

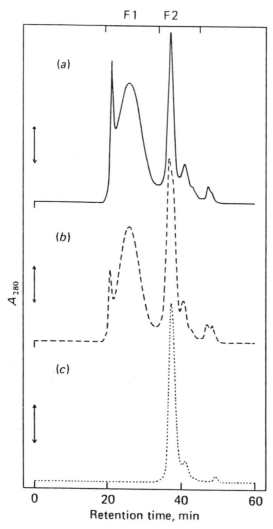

Figure 1. Elution patterns of casein micelles disaggregated in 6 M urea simulated milk ultrafiltrate. (a), Casein micelles; (b), reduced casein micelles; (c), reduced colloidal phosphate-free micelles. Flow rate was 0.5 ml/min. (Reproduced with permission from Ref. 13. Copyright 1986 Cambridge University Press.)

The proportion of fraction 1 decreased from 67.5 to 57.3% on reduction of casein micelles. This decrease was considered to be caused by dissociation of κ-casein since the decrease of fraction 1 amounted to about 10% of whole acid casein.

Fractions 1 and 2 of reduced casein micelles were prepared in larger quantities for further study. No changes in the separation ratio of fractions 1 and 2 were observed though the large peaks became broader and the small peaks appeared as shoulders rather than discrete peaks. Fractions 1 and 2 were concentrated by ultrafiltration and the amounts of calcium and inorganic phosphorus bound to casein were determined. Fraction 1 contained 1.7 times as much as calcium and inorganic phosphorus as did whole casein micelles, while fraction 2 contained a slight amount of bound inorganic phosphorus and less bound calcium than did whole casein micelles (Table I). Slight changes in the binding of calcium and CCP to casein might occur during the high performance gel chromatography and concentration by ultrafiltration. The presence of a slight amount of bound inorganic phosphorus in fraction 2 was probably due to contamination of fraction 1, since on rechromatography a small amount of fraction 1 was apparent in fraction 2. These facts indicate that fraction 1 of reduced casein micelles consisted of casein aggregates cross-linked by CCP while fraction 2 consisted of monomers of individual casein constituents binding only calcium.

Table I. Amounts of Ca and Inorganic P Bound to
Casein in Fractions 1 and 2

Sample	Bound Ca (mol/10 g casein)	Bound P (mol/10g casein)
Fraction 1	15.13	6.96
Fraction 2	3.78	0.58
Whole micelles	8.68	4.03

SOURCE: Reprinted with permission from ref. 13. Copyright 1986 Cambridge University Press.

Casein micelles are formed artificially when calcium alone is added to sodium caseinate solution. Such artificial casein micelles differ from natural ones in such properties as stability toward dialysis and toward pressure (14). Calcium alone does not form intermolecular cross-linkage since calcium caseinate dissociates to monomers in 4 M urea (15). It has been proposed that CCP links together submicelles between CCP (1). However, the existence of linkage between CCP and casein has not been proved. In the present study, the casein aggregates cross-linked by CCP were separated from reduced casein micelles by means of high performance gel chromatography on a TSK-GEL G4000SW column using USMUF as the effluent.

Chen and Yamauchi (16) and Morr (17) reported that casein micelles were separated into two fractions in 6.6 M urea and that the proportion of the fraction having larger molecular weight was 50% on a Sephadex G-200 column and 41% by analytical ultracentrifuge, respectively. The corresponding value in the present study was 67.5%. This discrepancy is primary due to partial dissociation of casein aggregates cross-linked by CCP in a urea solution.

It is probable that CCP is solubilized in a urea solution, though its solubilization is prevented in USMUF.

To examine the casein composition of fractions 1 and 2, high performance ion-exchange chromatography on a TSK-GEL DEAE 5PW column was performed (18). Eluted proteins were divided into four fractions, α_{s1}-, α_{s2}-, β- and κ-casein. Quantitative determination of casein constituents was made by peak area measurement of the chromatogram combined with individual extinction coefficients at 280 nm reported by Swaisgood (19). Table II shows the estimated relative composition for α_{s1}-, α_{s2}-, β- and κ-casein. The relative composition of whole casein agrees with that reported by Davies and Law (20), who used DEAE-cellulose column chromatography and the whole casein of bulk milk from Ayrshire cows. Fraction 1 contained 1.22 times as much as α_{s1}-casein and 1.61 as much as α_{s2}-casein as did whole micellar casein. On the other hand, fraction 2 contained 1.31 as much β-casein and 2.14 as much κ-casein as did whole micellar casein.

Table II. Casein Composition of Fractions 1 and 2

Sample	α_{s1} (%)	α_{s2} (%)	β (%)	κ (%)
Fraction 1	49.3	16.6	34.1	0
Fraction 2	23.7	3.5	50.8	22.0
Whole micelle	40.5	10.3	38.9	10.3

Dissociation of Casein Aggregates Cross-linked by CCP during Dialysis

Casein micelles are disaggregated by removal of calcium with calcium-chelating agents (8,21-23) or by dialysis against calcium-free buffer (24-26). Dialysis of skim milk against calcium-free buffer progressively solubilizes CCP while addition of calcium-chelating agent initially causes drastic changes in the ionic environment of casein micelles. Accordingly, dialysis has been utilized to clarify structure of casein micelles. When casein micelle dispersion (CMD), which was prepared by dispersing ultracentrifuged casein micelles in simulated milk ultrafiltrate, was dialyzed against 10 mM imidazole buffer (pH 7.0) at 5°C, casein micelles disaggregated (27). The changes in the micellar casein content are shown in Figure 2. Since the volume of CMD changed during dialysis, a correction was made on the basis of the total casein content which was determined on the dialyzed CMD. Micellar casein content is presented as a percentage of the total casein in the original CMD. The micellar casein content decreased to 11% after dialysis for 72 h. The colloidal calcium and inorganic phosphorus were also determined. The amounts of calcium and inorganic phosphorus decreased respectively from 77 to 11 mg and from 31 to 2 mg in 100 ml during dialysis for 72 h.

High performance gel chromatography of the casein micelles separated from the dialyzed CMD was carried out to determined the content of casein aggregates cross-linked by CCP. The elution patterns of casein micelles were shown in Figure 3. Fraction 1

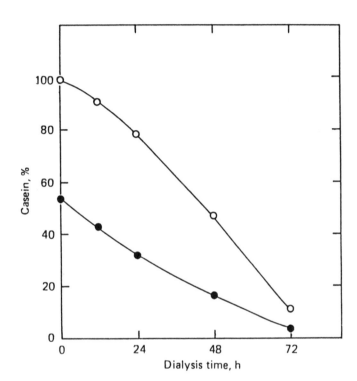

Figure 2. Changes in content of micellar casein (○) and casein aggregates cross-linked by CCP (●). (Reproduced with permission from Ref. 27. Copyright 1988 Cambridge University Press.)

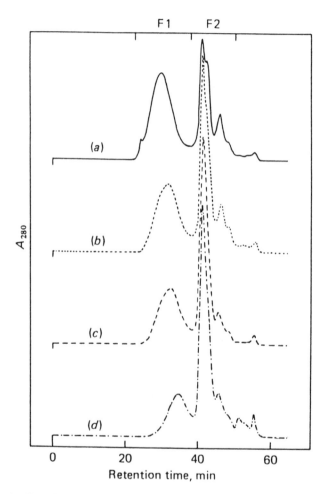

Figure 3. Elution patterns of casein micelles disaggregated in 6M urea simulated milk ultrafiltrate.   Casein   micelles   were obtained after 0 h (a), 24 h (b), 48 h (c) and   72   h   (d)   of dialysis. Flow rate was 0.5 ml/min.

(Reproduced with permission from ref. 37. Copyright 1989 Cambridge University Press.)

amounted 55.3% of the original micelles, and 44.4, 39.2 and 28.8% of the micelles obtained after 24, 48 and 72 h of dialysis, respectively. The percentage of casein aggregates cross-linked by CCP to total casein of CMD was calculated from the micellar casein content and fraction 1 content. The casein cross-linked by CCP decreased to 3% after 72 h dialysis (Figure 2). The retention time of the peak of fraction 1 was prolonged from 29.8 to 35.2 min by 72 h of dialysis.

The casein composition of casein aggregates cross-linked by CCP was determined by high performance ion-exchange chromatography. As described above, casein aggregates cross-linked by CCP in the undialyzed micelles contained more α_{s1}-, α_{s2}- and less β-casein than whole casein, and no κ-casein. The relative amount of α_{s2}-casein increased during dialysis while the relative amount of β-casein decreased (Table III). The casein aggregates cross-linked by CCP in the micelles obtained after 72 h dialysis contained 3.6 times as much α_{s2}-casein as did whole casein and less than half of the β- casein of whole casein. Only small changes in the relative amount of α_{s1}-casein were observed during dialysis.

Table III.   Casein Composition of Fraction 1 of Dialyzed
Casein Micelles

Dialysis time (h)	α_{s1} (%)	α_{s2} (%)	β (%)	κ (%)
0	50.3	14.6	35.1	0
24	51.6	18.6	29.8	0
48	50.4	28.7	20.9	0
72	48.5	36.0	15.5	0
Whole micelles	39.4	10.1	38.7	11.8

The amount of individual casein constituents cross-linked by CCP were calculated from fraction 1 content and casein composition. Dissociation of the individual casein constituents from casein aggregates cross-linked by CCP during dialysis is illustrated in Figure 4. In order to compare dissociation rates · of individual casein constituents, the percentage of each individual casein constituent cross-linked by CCP in the dialyzed micelles to that cross-linked by CCP in the original micelles has been indicated as the ordinate. The dissociation rates of individual casein constituents from casein aggregates cross-linked by CCP was in the order β- > α_{s1}- > α_{s2}-casein. α_{s1}-, α_{s2}- and β-Casein contain 8-9, 10-13 and 5 ester phosphate groups, respectively. There is only one ester phosphate group in κ-casein. The order of dissociation of individual casein constituents corresponds to that of the ester phosphate group content. These facts suggest that the higher the ester phosphate content, the stronger is interaction between CCP and casein molecules. In bovine casein micelles, ester phosphate-rich casein constituents such as α_{s2}-casein are considered to be cross-linked more tightly by CCP.

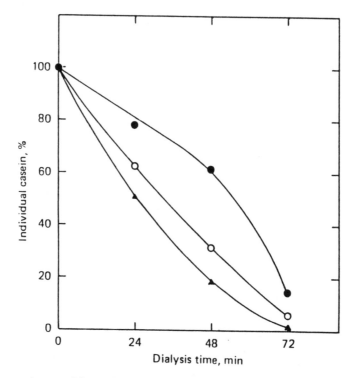

Figure 4.   Dissociation of individual casein constituents
from casein aggregates cross-linked by CCP during dialysis.
○ , α_{s1}-Casein; ●, α_{s2}-casein; ▲, β-casein.   (Reproduced
with permission from Ref. 27. Copyright 1988 Cambridge University Press.)

Role of Individual Milk Salt Constituents in Cross-linking by CCP

Artificial casein micelles were prepared at a casein concentration of 2.5% by adding various salt solutions to sodium caseinate solution. Although casein micelles were formed artificially when calcium chloride alone was added to a sodium caseinate solution, calcium alone did not form an intermolecular cross-linkage. When phosphate was incorporated into the calcium caseinate micelle systems, fraction 1 appeared (28). Although fraction 1 increased with increasing calcium and phosphate concentrations up to 30 mM calcium and 22 mM phosphate, it decreased at higher calcium and phosphate concentrations (Table IV). The amounts of colloidal calcium and phosphate, which were separated by ultracentrifugation, in the artificial casein micelles formed by calcium and phosphate increased with increasing calcium and phosphate concentrations. However, part of the colloidal calcium and phosphate was insoluble in the urea solution, suggesting that urea-insoluble calcium phosphate does not participate in cross-linking the casein molecules and exists in the separated phase. In samples 3 and 4, 6.7 and 9.7% of casein were not solubilized by 6 M urea. This urea-insoluble casein is considered to have been adsorbed by urea-insoluble calcium phosphate, as in the case of a hydroxyapatite column (29). When casein micelles were formed by calcium, phosphate and citrate, the amount of colloidal calcium and phosphate were smaller than when prepared in the absence of citrate. In sample 5, almost no CCP was formed and fraction 1 was not observed. However, at higher calcium and phosphate concentrations, the fraction 1 content was higher in micelles containing citrate than in micelles containing no citrate. When citrate was increased from 10 to 15 mM, the fraction 1 content decreased from 46.6 to 36.2%.

Table IV. Distribution of Salts and Fraction 1 Content
in Artificial Casein Micelles

Sample	Total			Colloidal		Urea-insoluble		
no.	Ca (mM)	P (mM)	Citrate (mM)	Ca (mM)	P (mM)	Ca (mM)	P (mM)	F1 (%)
1	10	7.3	0	4.7	1.9	0	0	7.2
2	20	14.6	0	18.5	8.4	5.2	2.3	28.4
3	30	22.0	0	27.0	15.4	8.3	5.7	43.1
4	40	29.3	0	35.3	22.5	19.3	13.5	40.9
5	10	7.3	10	0.5	0.1	0	0	0
6	20	14.6	10	6.5	3.3	0	0	13.2
7	30	22.0	10	19.7	9.8	0	0	46.6
8	40	29.3	10	27.2	16.0	6.2	3.9	54.1
9	30	22.0	15	17.7	7.2	0	0	36.2

In order to clarify the role of citrate in cross-linking by CCP, additional experiments were carried out. The precipitate of calcium phosphate was previously formed in the presence and absence of citrate, and then casein solution was added to the suspensions. After these solutions had been stirred for 4 and 8 h, solid urea was added. Fraction 1 was formed in the presence of citrate although its content was lower than that of sample 7, while fraction 1 was not formed in the absence of citrate. Magnesium forms a precipitate with phosphate as does calcium. However, when magnesium, phosphate and citrate were added to the casein solution, fraction 1 was not observed. This suggests that magnesium phosphate has no cross-linking ability. When 5 mM magnesium was added to the micelle system formed by 30 mM calcium, 22 mM phosphate and 10 mM citrate, the amount of colloidal calcium and phosphate and the content of fraction 1 increased.

Schmidt (1) has proposed that the initial precipitate which forms when calcium- and phosphate-containing solutions are mixed is invariably amorphous calcium phosphate (ACP); that ACP is unstable and is transformed into stable crystalline hydroxyapatite (HAP) after a certain induction time; that casein and citrate inhibit the transformation; and that the CCP in milk is ACP. In fact casein and phosphopeptides inhibit the precipitation of calcium phosphate (30,31). Recently, van Kemenade (32) reported that caseins inhibited the growth of brushite, octacalcium phosphate and HAP by binding to them through their ester phosphate groups. Calcium phosphate formed in the absence of citrate may be transformed from ACP to HAP. This may explain the reason why the precipitate of calcium phosphate formed in the absence of citrate has no cross-linking ability, if ACP and not HAP has no cross-linking ability.

Both magnesium and calcium are alkaline earth-metals and bind to casein. Calcium forms stable α_{s1}-κ-casein micelles, but magnesium does not (33). Magnesium is a constituent of natural casein micelles and is contained in a calcium phosphate citrate complex (34). Magnesium phosphate alone had no cross-linking ability. However, magnesium increased the amount of calcium phosphate to promote cross-linking.

Schmidt et al (14,35,36) have studied the artificial casein micelles in detail. They showed that calcium caseinate micelles differed from the natural micelles in their size distribution and behavior, particularly with respect to their stability toward dialysis and high static pressures. By incorporating magnesium, phosphate and citrate into calcium caseinate micelle systems, they prepared artificial casein micelles which showed the same behavior as natural micelles. The intermolecular cross-linking by CCP is considered to be responsible for the characteristic behavior of natural micelles.

Incorporation of Individual Casein Constituents into Casein Aggregates Cross-linked by CCP

Bovine caseins are synthesized in the mammary gland and micelle formation occurs in the Golgi vesicles. It is assumed that casein submicelles are aggregated into micelles by CCP formed by transformation of calcium and other inorganic constituents from cytosol to the Golgi vesicles (1,3). However, the formation of cross-linkages between casein molecules and CCP in the bioassembly of casein micelles has not yet been elucidated sufficiently. In order to examine the process of cross-linking of casein by CCP, artificial casein micelles were prepared at a casein concentration of 2.5% with 10-40 mM calcium, 12-27 mM phosphate and 10 mM citrate (37). No micelles were formed at 10 mM calcium, 12 mM phosphate and 10 mM citrate, which are the approximate concentrations found in the soluble phase of bovine milk. The amounts of micelles formed by addition of calcium and phosphate increased with increasing calcium and phosphate concentrations: the micellar casein content was 56% and 95% in the samples formed at 17.5 mM calcium 15.75 mM phosphate and 30 mM calcium 22 mM phosphate, respectively. The high performance gel chromatography of casein micelles sedimented by ultracentrifugation was carried out in order to determine the content of casein aggregates cross-linked by CCP. The casein micelles formed at low calcium and phosphate concentrations were low in the content of casein aggregates cross-linked by CCP. The retention time for the peak of fraction 1 was shortened with increasing calcium and phosphate concentrations. The percentage of casein aggregates cross-linked by CCP to total casein was calculated from the micellar casein content and the content of casein aggregates cross-linked by CCP (Figure 5). The concentration of colloidal phosphate is also shown in Figure 5. Although both the casein aggregates cross-linked by CCP and the colloidal phosphate increased with increasing calcium and phosphate concentrations, they differed significantly from each other in this respect. This may be due to changes in the molecular weight of casein aggregates cross-linked by CCP, which results in changes in the retention time of fraction 1.

The composition of casein aggregates cross-linked by CCP was determined by high-performance gel chromatography. Casein aggregates cross-linked by CCP contained more α_{s1}-, α_{s2}- and less β-casein than whole casein. The relative amount of α_{s2}-casein was higher in the samples with lower calcium and phosphate concentrations, while the relative amount of β-casein was lower. There were only small changes in the relative amount of α_{s1}-casein. The amount of individual casein constituents cross-linked by CCP were calculated from the content of casein aggregates cross-linked by CCP. Incorporation of individual casein constituents into casein aggregates cross-linked by CCP is illustrated in Figure 6. In order to compare the incorporation rates into casein aggregates cross-linked by CCP, the percentage of each individual casein constituent cross-linked by CCP to the total content is indicated at the ordinate and the percentage of casein aggregates cross-linked by CCP to the total casein is indicated at the abscissa. The incorporation rates of individual casein constituents into casein aggregates cross-linked by CCP were in the order α_{s2}- > α_{s1}- > β-casein. This is the reverse of the order of the dissociation rates of casein aggregates cross-linked by CCP during dialysis.

The higher the ester phosphate content, the higher the incorporation rate of the individual casein constituent and slower is the dissociation rate. It is suggested that initially submicelles rich in ester phosphate groups, i.e. rich in α_{s1}- or α_{s2}-casein, are bound together through ester phosphate groups by CCP.

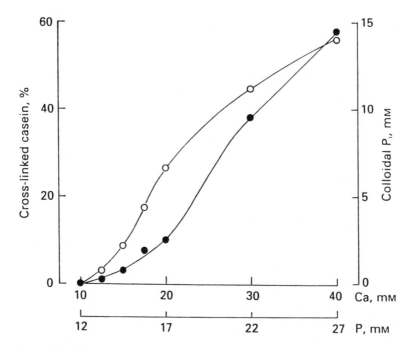

Figure 5. Changes in content of casein aggregates cross-linked
by CCP and colloidal inorganic phosphate (Pi) by addition of
calcium and phosphate. O , Cross-linked casein ; ● , Pi.
(Reproduced with permission from Ref. 37.   Copyright 1989
Cambridge University Press.)

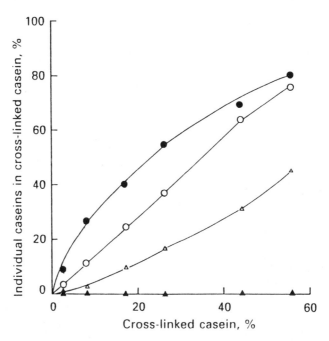

Figure 6. Partition of individual casein constituents in casein aggregates cross-linked by CCP. $\bigcirc$, α_{s1}-Casein; $\bullet$, α_{s2}-casein; $\triangle$, β-casein; $\blacktriangle$, κ-casein. (Reproduced with permission from Ref. 37.   Copyright 1989 Cambridge University Press.)

The Sites of Casein for Interaction with CCP

McMahon and Brown (6) have proposed that both carboxyl and ester phosphate groups are possible sites for linkages between casein and CCP. Schmidt (1) has presumed that ester phosphate groups of casein are potential sites for interaction with CCP. However, the identity of the sites for interaction with CCP has not yet been proved completely. In order to identify the sites of casein for interaction with CCP, artificial casein micelles were prepared with α_{s1}-κ- and β-κ-caseins, the weight ratio of which was 9 : 1, at 30 mM calcium, 22 mM phosphate and 10 mM citrate (38). Fraction 1 contents of α_{s1}-κ- and β-κ-casein micelles were $\overline{53.9}$ and 31.8%, respectively. When calcium, phosphate and citrate were added to κ- casein alone, no fraction 1 appeared (Table V), and calcium phosphate formed was observed as precipitates in the urea solution. When calcium, phosphate and citrate were added to the solution of γ-κ-caseins, again no fraction 1 was observed (Table V). This indicated that κ- and γ-casein were not cross-linked at all by CCP.

Table V. Fraction 1 Content of Artificial Casein
Micelles with Various Casein Combinations

Sample	Fraction 1 (%)
α_{s1}-κ-Casein micelles	53.9
β-κ-Caseinmicelles	31.8
κ-Casein micelles	0
γ-κ-Casein micelles	0
Dephosphorylated whole casein micelles	0
Dephosphorylated α_{s1}-casein micelles	0

In order to obtain further evidence for role of ester phosphate groups in cross-linking by CCP, enzymatically dephosphorylated whole casein and α_{s1}-casein were prepared. The phosphorus content of dephosphorylated whole casein and α_{s1}-casein were 0.058% and 0.031%, respectively. The average ester phosphate content of dephosphorylated α_{s1}-casein was estimated to be 0.24 per molecule. Dephosphorylated α_{s1}-casein forms micelles in the presence of calcium and κ-casein (39). However, neither in dephosphorylated whole casein micelles nor in dephosphorylated α_{s1}-κ-casein micelles was any fraction 1 formed (Table V).
    Since κ-casein was not cross-linked by CCP, it was chemically phosphorylated by the method of Ullman and Perlman (40). The average phosphate content of chemically phosphorylated $\overline{\kappa}$-casein was estimated to be 8.5 per molecule. The prepared chemically phosphorylated κ-casein contained a slight amount of the covalently cross-linked components formed by side reactions during the phosphorylation process. When 30 mM calcium, 22 mM phosphate and 10 mM citrate were added to the chemically phosphorylated κ-casein solution, cross-linked components

appeared. Accordingly, it is concluded that the sites for interaction with CCP are ester phosphate groups.

Relation between CCP Cross-linkage and Changes in Casein Micelles on Heating and Cooling

Heating milk causes a number of changes to occur within casein micelles. It is well known that casein micelles aggregate on heating milk at high temperature (41). However, heating milk above 105°C also causes disaggregation of casein micelles with formation of serum casein which is accelerated by concentration of milk (42,43). A larger amount of serum casein is formed on heating from smaller micelles than from larger micelles (44). There is only a slight effect of CCP level on formation of serum casein on heating. The dependence of micelle size for the formation of serum casein on heating is ascribed to be variation in κ-casein content rather than CCP content. There is some evidence of a possible change in the state of CCP of heated milk (3). The amount of serum casein formed on the cooling of the casein micelles which had been heated at 135-140°C for 15 s is much larger than formed on cooling unheated micelles (45). This suggests that the interaction between CCP and casein is weakened on heating milk at high temperature. When CMD was heated at 135-140°C for 15 s, the content of fraction 1 decreased from 51.9 to 46.1%, indicating the cleavage of CCP cross-linkage. The cleavage of CCP linkage was considered to occur without liberation of ester phosphate groups. It was suggested that the transformation of CCP to another form was responsible for the cleavage of CCP cross-linkage.

It is well known that serum casein is released from casein micelles (8-11). This is related to the temperature dependent dissociation of β-casein and the structure of casein micelles since the major component of the serum casein is β-casein. Since about half of β-casein is cross-linked in casein micelles by CCP, it is probable that the β-casein released on cooling arises from β-casein which is not cross-linked. It is expected that the release of serum casein is depressed when the content of fraction 1 is increased. Accordingly, CCP-increased CMD was prepared by addition of calcium and phosphate. The amount of serum casein released from casein micelles decreased with increasing CCP content.

Conclusion

CCP is an essential constituent and an integral factor in casein micelles, since removal of it disaggregates casein micelles to submicelles. By providing direct experimental evidence for the existence of the linkage between casein and CCP, we have separated casein aggregates cross-linked by CCP from bovine casein micelles by high high performance gel chromatography in the presence of urea. Caseins are cross-linked through their ester phosphate groups by CCP. The higher the ester phosphate content of casein, the stronger is the interaction between CCP and casein molecules.

Riboflavin binding protein from hen egg white, which consists 219 amino acid residues and contains 8 ester phosphate groups between Ser (187) and Ser (197), can be cross-linked by calcium phosphate (unpublished results). Although κ-casein is not cross-linked by CCP, chemically phosphorylated κ-casein is cross-linked. Phosphorylation of food proteins improves functional properties such as gel-forming properties and foaming properties. It is expected that cross-linking of phosphoproteins by calcium phosphate further improves their functional properties. It is now recognized that casein micelles act as carriers of calcium phosphate, since phosphopeptides from casein promote absorption of calcium in the intestine (46). Phosphoproteins cross-linked by calcium phosphate may possibly possess similar functional properties to those of casein micelles. Accordingly, it is expected that cross-linking of phosphoproteins by calcium phosphate improves not only functional properties but also nutritive values.

Literature Cited

1. Schmidt, D.G. In Developments in Dairy Chemistry Vol. 1, Proteins; Fox, P.F., Ed.; Applied Science Publishers Ltd.: London, 1982; p 61.
2. Pyne, G.T. J. Dairy Res. 1962, 29, 101.
3. Holt, C. In Developments in Dairy Chemistry Vol. 3, Lactose and Minor Constituents; Fox, P.F., Ed.; Applied Science Publishers Ltd.: London, 1985; p 143.
4. Morr, C.V.; Josephson, R.V.; Jenness, R.; Manning, P.B. J. Dairy Sci. 1971, 54, 1555.
5. Slattery, C.W. J. Dairy Sci. 1976, 59, 1547.
6. McMahon, D.J.; Brown, R.J. J. Dairy Sci. 1984, 67, 499.
7. Fox, P.F. In Developments in Dairy Chemistry Vol. 1, Proteins; Fox, P.F., Ed.; Applied Science Publishers Ltd.: London, 1982; p 189.
8. Rose, D. J. Dairy Sci. 1968, 51, 1897.
9. Downey, W.K.; Murphy, R.F. J. Dairy Res. 1970, 37, 361.
10. Ali, A.E.; Andrews, A.T.; Cheeseman, G.C. J. Dairy Res. 1980, 47, 371.
11. Davies, D.T.; Law, A.J.R. J. Dairy Res. 1983, 50, 67.
12. McGann, T.C.A.; Fox, P.F. J. Dairy Res. 1974, 41, 45.
13. Aoki, T.; Kako, Y.; Imamura, T. J. Dairy Res. 1986, 53, 53.
14. Schmidt, D.G.; Koops, J.; Westerbeek, D. Neth. Milk Dairy J. 1977, 31, 328.
15. Ono, T.; Kaminogawa, S.; Odagiri, S.; Yamauchi, K. Agric. Biol. Chem. 1976, 40, 1725.
16. Chen, C.M.; Yamauchi, K. Agric. Biol. Chem. 1971, 35, 637.
17. Morr, C.V. J. Dairy Sci. 1967, 50, 1744.
18. Aoki, T.; Yamada, N.; Kako, Y.; Kuwata, T.; Jpn J. Zootech. Sci. 1986, 57, 624.
19. Swaisgood, H.E. In Developments Dairy Chemistry Vol. 1. Proteins; Fox, P.F. Ed.; Applied Science Publishers Ltd.: London, 1982; p 1.
20. Davies, D.T.; Law, A.J.R. J. Dairy Res. 1977, 44, 213.

21. Noble, R.W.; Waugh, D.F. J. Am. Chem. Soc. 1965. 87, 2236.
22. Lin, S.H.C.; Leong, S.L.; Dewan, R.K.; Bloomfield, V.A.;
    Morr, C.V. Biochemistry 1972, 11, 1818.
23. Ono, T.; Hayakawa, Y.; Odagiri, S. J. Agric. Chem. Soc.
    Jpn. 1975, 49, 417.
24. Schmidt, D.G.; Buchheim, W. Milchwissenschaft 1970, 25,
    596.
25. Ono, T.; Dan, H.T.; Odagiri, S. Agric. Biol. Chem. 1978,
    42,
    1063.
26. Ono, T.; Furuyama, T.; Odagiri, S. Agric. Biol. Chem.
    1981, 45, 511.
27. Aoki, T.; Yamada, N.; Kako, Y. J. Dairy Res. 1988, 55, 189.
28. Aoki, T.; Kawahara, A.; Kako, Y.; Imamura, T. Agric. Biol.
    Chem. 1987, 51, 817.
29. Addeo, F.; Chobert, J.M.; Ribadeau-Dumas, B. J. Dairy Res.
    1977, 44, 63.
30. Schmidt, D.G.; Both, P. Neth. Milk Dairy J. 1987, 41, 105.
31. Schmidt, D.G.; Both, P.; Visser, S.; Slangen, K.J.; van
    Rooijen, P.J. Neth. Milk Dairy J. 1987, 41, 121.
32. van Kemenade, M.J.J.M.; de Bruyn, P.L. J. Colloid Interface
    Sci. 1987, 118, 564.
33. Kaminogawa, S.; Koide, K.; Yamauchi, K. Agric. Biol. Chem.
    1977, 41, 697.
34. McGann, T.C.A.; Kearney, R.D.; Buchheim, W.; Posner, A.S.;
    Betts, F.; Blumennthal, N.C. Calcified Tissue Int.
    1983, 35, 821.
35. Schmidt, D.G.; Koops, J. Neth. Milk Dairy J. 1977, 31, 342.
36. Schmidt, D.G.; Both, P.; Koops, J. Neth. Milk Dairy J.
    1979, 33, 40.
37. Aoki, T. J. Dairy Res. 1989, 56, 613.
38. Aoki, T.; Yamada, N.; Tomita, I.; Kako, Y.; Imamura, T.
    Biochim. Biophys. Acta 1987, 911, 238.
39. Bingham, E.W.; Farrell, H.M,Jr.; Carroll, R.J. Biochemsitry
    1973, 11, 2450.
40. Ullman, B.; Perlman, R.L. Biochem. Biophys. Res. Commun.
    1975, 63, 424.
41. Fox, P.F. J. Dairy Sci. 1981, 64, 2127.
42. Aoki, T.; Suzuki, H.; Imamura, T. Milchwissenschaft, 1974,
    29, 589.
43. Aoki, T.; Suzuki, H.; Imamura, T. Milchwissenschaft, 1975,
    30, 30.
44. Aoki, T.; Kako, Y. J. Dairy Res., 1983, 50, 207.
45. Aoki, T.; Imamura, T. Agric. Biol. Chem., 1975, 39, 2107.
46. Lee, Y.S.; Noguchi, T.; Naito, H. Agric. Biol. Chem. 1979,
    43, 2009.

RECEIVED August 14, 1990

Chapter 13

Quaternary Structural Changes of Bovine Casein by Small-Angle X-ray Scattering

Effect of Genetic Variation

T. F. Kumosinski, H. Pessen, E. M. Brown, L. T. Kakalis, and H. M. Farrell, Jr.

Eastern Regional Research Center, Agricultural Research Service, U.S. Department of Agriculture, Philadelphia, PA 19118

Milks containing the A or B genetic variants of α_{s1}-casein have markedly different physical properties (solubility, heat stability). When examined by small-angle X-ray scattering (SAXS), whole caseins, either A or B, as submicelles (without Ca^{2+}) behaved as inhomogeneous spheres with two concentric regions; the inner (more electron dense) region displayed protein-protein inter- actions, the outer region, high hydration. Upon addition of Ca^{2+}, casein of both variants, while retaining its submicellar hydration and structure was packed into colloidal micelles at ratios of 3:1 for B, and 6:1 for A. Tightly bound water (by ^2H NMR relaxation) was only a fraction of total water (by SAXS): thus both micelles and submicelles contained trapped water. For both micelles and submicelles, similar dynamic motions were observed by ^{13}C NMR. Open penetrable 3-D structures of α_{s1}- and κ-caseins were predicted by energy minimization. All results support a model featuring micelles composed of submicelles which exhibit high mobility accounting for the known diffusion of enzymes and cosolutes throughout the casein micelle.

Whole casein occurs in bovine milk as a colloidal calcium-phosphate-containing protein complex, commonly called the casein micelle. The micellar structure is disrupted by the removal of calcium, resulting in noncolloidal protein complexes called submicelles (1). These submicelles consist of four proteins, α_{s1}-, α_{s2}-, β- and κ-casein, in the approximate ratios of 4:1:4:1 (2). All are phosphorylated to various extents and have an average monomer molecular weight of 23,000 (3). Isolated casein fractions exhibit varying degrees of self-association which are mostly hydro- phobically driven (1). The nature of tertiary and quaternary structures of native caseins in mixed association has received little attention. However, there is hydrodynamic evidence that, in the absence of calcium, casein monomers associate to form aggregates, submicelles, with a maximum Stokes radius of 9.4 nm (4).

Upon the addition of calcium, these hydrophobically stabilized casein submicelles further self-associate to colloidal micelles with average radii of 65 nm. The formation of micelles is thought to occur via calcium-protein side-chain salt bridges (1,4). The exact supramolecular structure of the casein micelle remains a topic of controversy. Proposed models of the casein micelle include: a micelle composed of discrete submicelles (3); a loose, porous gel structure (5); and a newer model of a homogeneous sphere with a "hairy" outer layer (6).

X-ray crystallography is generally the technique of choice for the elucidation of protein structure. However, for proteins such as casein, which do not crystallize, valuable molecular information may be extracted by the companion technique of small-angle X-ray scattering

(SAXS). This technique which measures the intensity of scattering produced by the electrons of the solute can yield valuable information including degree of hydration, radius of gyration and molecular weight (7).

SAXS was undertaken on whole bovine casein from two distinct genetic lines (α_{s1}-caseins A and B), first in the absence of calcium to describe the nature of the limiting polymer structure of these two whole caseins (submicellar structures) and second, in the presence of calcium, to determine if the colloidal micelle consists of discrete submicellar particles with a particular packing structure or of a nonspecific, unordered, gel-like structure. Since milks containing these two genetic variants differ in their physical properties (8), information on the molecular basis for these differences could be assessed as well. The results of these studies are correlated with previous studies on micelles and submicelles using ^{2}H NMR relaxation and ^{13}C NMR spectroscopy of the caseins.

Materials and Methods

Sample Preparation. Whole sodium caseinate from the milks of two individually selected cows was prepared as described previously (9). These caseins were of the genotypes (α_{s1}-AA, β-AA, κ-AA) and (α_{s1}-BB, β-AA, κ-BB) (10). Casein micelles were prepared by the addition of CaCl$_2$ (final concentration 10 mM) to a solution of lyophilized caseinate in PIPES-KCl buffer (25 mM piperazine-N-N'-bis (2-ethanesulfonic acid), pH 6.75, made to be 80mM in KCl) (9). For the preparation of submicelles the CaCl$_2$ was replaced by additional KCl (30 mM to match the ionic strength of the CaCl$_2$).

SAXS Measurements and Data Analysis. The measurement of SAXS and evaluation of data were as described previously (9) using the Cu-K$_\alpha$ doublet at 0.154 nm. For data evaluation, the partial specific volume $\bar{v}$ and number of electrons per gram of particle were calculated from the amino acid composition (11). The computer program of Lake (12) was used to deconvolute slit-smeared curves. All data were fitted to multiple Gaussian functions by the use of a Gauss-Newton nonlinear-regression computer program developed at this laboratory (9). Lowest root mean square variation and random residuals were used as criteria for the number of Gaussians used.

NMR Measurements. Proton-decoupled, natural abundance ^{13}C (100.5 MHz) NMR measurements were carried out with a JEOL GX-400 multinuclear spectrometer. (Reference to brand or firm name here and in the following does not constitute endorsement by the U.S. Department of Agriculture over others of a similar nature not mentioned.) Generally, 240 mg of lyophilized caseinate containing α_{s1}-B were dissolved in 4 ml D$_2$O (6.0 % w/v), containing KCl for casein submicelles or KCl-CaCl$_2$ for casein micelles, together with 0.3 mg/ml sodium 2,2-dimethyl-2-silapentane-5 sulfonate (DSS) as an internal chemical shift standard. The 90^0 pulsewidth was 24 ms, the spectral width 25 kHz, the acquisition time 0.65 s and a 32K point time-domain array was used for storing the data. T$_1$ values were measured by inversion-recovery (13). Approximately 30 min were allowed for each sample to reach thermal equilibrium in the magnet before data acquisition. The probe temperature was controlled ($\pm$ 0.5^{0}C) by means of a thermostated dry nitrogen current.

Molecular Modeling. Three-dimensional (3D) representations of casein monomers were constructed using Sybyl-Mendel molecular modeling programs with Evans and Sutherland hardware. Selection of appropriate conformation states for the individual amino acid residues was accomplished by comparing the results of sequence-based predictive techniques (14,15) with available spectroscopic data (16,17).

Results and Discussion

Submicelles. The SAXS data were analyzed by nonlinear regression and fitted by the sum of two Gaussian functions. For both genetic variants (A and B), these data were interpreted by means of a model in which the particle has two regions of different electron densities with the same scattering center. In this model, the scattered amplitudes rather than the intensities of the two regions must be added because of interference effects of the scattered radiation. Molecular and structural parameters for the two caseins under submicellar conditions were evaluated using equations and notation developed by Luzzati et al. ([18,19]). These parameters are listed in Tables I and II where subscripts C and L refer to parameters for the higher (core) and lower (shell or loose) electron density regions, respectively, while subscript 2 refers to the particle including both regions.

Variant B. The molecular weight, M_2, found for the total submicellar particle containing α_{s1}-B casein, the more prevalent variant, was 285,000 ± 14,600. Both M_2 and k, the mass fraction of the denser or "core" region, were independent of protein concentration, ruling out an explanation of the multiple Gaussian character of the scattering as due to extreme particle size polydispersity ([20]). Extreme particle asymmetry (e.g., rods) can be ruled out from electron-microscopic, hydrodynamic, and light scattering evidence indicating approximately spherical particles ([1,21]). Hence the molecular parameters given in Table I are a measure of the limiting aggregate of the hydrophobically driven mixed self-association of the whole caseins in the absence of calcium. The molecular weight of this limiting polymer (M_2 = 285,000) is consistent with the 200,000 to 300,000 values found by a variety of techniques ([21,22]).

Table I. Molecular parameters of variants A and B (Desmeared SAXS)

Parameter	Submicelle		Micelle	
	A	B	A	B
M	-------	-------	2,090,000 ± 500,000	882,000 ± 28,000
k2	-------	-------	0.167 ± 0.038	0.308 ± 0.005
P #			6.0:1	3.2:1
M_2	312,000 ±19,000	285,000 ± 14,600	350,000 ± 28,000	276,000 ± 18,000
k	0.262 ± 0.009	0.212 ± 0.028	0.166 ± 0.010	0.216 ± 0.003
M_C	81,700 ± 4,600	60,000 ± 5,600	65,100 ± 1,900	56,400 ± 3,700
M_L	231,000 ± 15,000	225,000 ± 18,500	289,000 ± 30,000	220,000 ± 18,700
$\delta\rho$[a]	-------	-------	9.3± 0.9	8.0 ± 0.4
$\delta\rho_2$	10.7 ± 0.6	9.9 ± 0.4	8.4 ± 0.4	7.3 ± 0.5
$\delta\rho_C$	18.4 ± 0.8	14.8 ± 1.4	13.4 ± 0.2	12.8 ± 0.7
$\delta\rho_L$	8.1 ± 0.4	8.5 ± 0.3	7.8 ± 0.5	5.7 ± 0.3
H[b]	-------	-------	6.71 ± 0.63	7.92 ± 0.42
H_2	5.74 ± 0.27	6.31 ± 0.30	7.51 ± 0.34	8.98 ± 0.44
H_C	3.05 ± 0.17	3.97 ± 0.48	4.45 ± 0.08	4.70 ± 0.31
H_L	7.88 ± 0.40	7.41 ± 0.30	8.68 ± 0.56	11.44 ± 0.58

Values are averages for three concentrations. Units are [a] e^-/nm^3, and [b] $g_{H_2O}/g_{protein}$

In the small-angle neutron-scattering study of Stothart & Cebula ([22]), the data were analyzed on the basis of a model consisting of a homogeneous limiting aggregate. Here, we have found a heterogeneous particle consisting of two regions of differing electron density, with the mass fraction of the higher electron density region, k, equal to 0.212 ± 0.028. This core region, moreover, has an electron density difference, $\Delta\rho_C$ of 14.8 ± 1.4 e$^-$/nm^3, a hydration, H_C, of 3.97 ± 0.48 g water/g protein, and a molecular weight, M_C, of $60,000 \pm 5,600$ (Table I). The region of higher electron density most likely results from the intermolecular hydrophobically driven self-association of casein monomer units ([1]); the hydrophobic inner core would be surrounded by a less electron-dense region (a loose or shell area) presumably consisting mainly of hydrophilic groups ([23,24]). The hydration formally ascribed to the core region is likely to be a characteristic of the packing density ([25,26]) of the hydrophobic side chains rather than any actual amount of water "bound" within this region. Our own research ([27]) using ^{2}H NMR relaxation measurements in D$_2$O showed that isotropically bound water associated with submicelles occurs with an average rotational correlation time of 38 ns, thus yielding a Stokes radius of 3.6 nm (Table III). This value is in good agreement with R_C of Table II (3.8 nm) found by SAXS. Thus, the most tightly bound water may occur at the surface of this more electron dense inner core, while water occurring outside this limit may be considered trapped or protein-influenced. Indeed, the ratio of hydration in the loose region, H_L, is 1.7 times that of the core region, H_C, for the submicelles (Table II).

Table II. Structural parameters of variants A and B (Desmeared SAXS)

Parameter	Submicelle		Micelle	
	A	B	A	B
V, nm^3	-------	-------	26,080 ± 2,390	12,720 ± 250
V2, nm^3	3,400 ± 90	3,330 ± 260	4,880 ± 130	4,440 ± 160
VC, nm^3	495± 30	467 ± 2	519 ± 2	529 ± 3
Vl, nm^3	3,300 ± 60	3,320 ± 400	4,580 ± 90	4,310 ± 20
R$_2$, nm	8.51 ± 0.02	8.02 ± 0.04	8.99 ± 0.01	9.06 ± 0.01
R$_C$, nm	3.93 ± 0.01	3.80 ± 0.01	3.93 ± 0.01	3.96 ± 0.01
R$_L$, nm	9.62 ± 0.03	8.82 ± 0.08	9.69 ± 0.03	10.02 ± 0.01
R$_G$,* nm	6.28	7.72	27.22	17.52
Dmax,* nm	15.66	19.89	83.38	51.21
(a/b)C	1.46	1.33	-------	-------
(a/b)2	2.28	1.98	-------	-------

Values are averages for three concentrations. (*) - calculated from the distance distribution.

still low when compared to compact globular proteins ([20]), such as lysozyme (78 e$^-$/nm^3), α-lactalbumin (67 e$^-$/nm^3), ribonuclease (71 e$^-$/nm^3) or riboflavin-binding protein (56 e$^-$/nm^3). This result emphasizes the consequences of the unique nature of the conformation of the casein polypeptide chains. Caseins have long been regarded to have little secondary structure ([3]); however recent evidence from Raman spectroscopy ([16]) suggests that whole casein in the submicellar form may have more structure than estimated from the sum of the individual casein components. Moreover, the Raman data permit the estimation of the percentage of β-turns in whole casein. Nearly 40% of the casein structure occurs in β-turns, which demonstrates that the conformation of submicellar caseins is not that of a totally random structureless coil. This finding is further supported by recent 3D molecular modeling studies of bovine casein (28). Computer-generated 3D models of α_{s1}- and κ-caseins based on secondary structural predictions and Raman data showed a retention of β-turns during energy minimization. In addition, the overall shape and known biochemical properties (availability of proteolytic cleavage sites, disulfide geometry, and positions of phosphorylated sites) were in good agreement with published data. The 3D models are also consistent with the SAXS data since they show that the monomeric caseins have hydrophobic β-sheets which may associate to produce hydrophobic inner cores.

Variant A. The molecular weight, M_2, of the submicellar particle containing the A variant of α_{s1}-casein was 312,000 ± 19,000. Both M_2 and k, the mass fraction of the denser or "core" region, were invariant as a function of protein concentration. We find that that submicelles of α_{s1}-A, like those of the B variant, consist of two regions of differing electron density, with the mass fraction of the higher electron density region equal to 0.262 ± 0.009. This higher electron density region, moreover, has an electron density difference, $\Delta\rho_C$, of 18.4 ± 0.8 e^-/nm³, a hydration, H_C, of 3.05 ± 0.17 g water/g protein and a molecular weight, M_C, of 81,700 ± 4,600 (see Table I).

The derived structural parameters for the α_{s1}-A casein submicelles are listed in Table II. An axial ratio for the denser region, $(a/b)_C$, of 1.46 can be calculated from V_C and R_C (18), and a value of 2.28 for the axial ratio of the total submicelle, $(a/b)_2$, from V_2 and R_2, using as a model a prolate ellipsoid of revolution. These axial ratios, like the corresponding values of 1.33 and 1.98 for the B variant, are reasonable indications that the casein submicelle deviates only moderately from spherical symmetry, as would be predicted from electron microscopy (1).

Micelles. Whether the integrity of the submicellar structure is maintained within the colloidal micelle has been a subject of much controversy (5). To address this problem, the scattering of whole casein solutions of both genetic variants with 10 mM CaCl₂, but without phosphate buffer to compete with the protein calcium binding sites, was studied. The SAXS data for casein micelle solutions were fitted to the sum of three Gaussians. The two Gaussians having the smaller radii of gyration constitute the contribution of the submicellar structure to the SAXS results. The third Gaussian, which has the largest radius of gyration, reflects the total number of submicellar particles within the cross-sectional SAXS scattering profile. Here, at zero angle, the intensity of the larger Gaussian contribution can be simply added to the intensity of submicellar contribution. A new parameter, k_2, the ratio of the mass of the submicelles to the total observed mass ascribable to a cross section, can be expressed in terms of the radii of gyration and the zero-angle intercepts for the three Gaussians. The packing number, the reciprocal of k_2, is the number of submicellar particles found within a micellar cross section. The meaning to be ascribed to the cross section in this context will be discussed further below. The resulting parameters for micelles of both genetic variants are listed in columns 4 and 5 of Tables I and II, where subscript 2 now designates the corresponding parameters for a submicellar particle when incorporated in the micelle, and unsubscripted parameters refer to the total cross section of the colloidal particle.

Variant B. As seen in Table I, k_2 for casein micelles of α_{s1}-B was 0.308 ± 0.005, and the packing number, its reciprocal, was 3.2. The large average radius of the micelles (65 nm) implies corresponding scattering angles too small to be experimentally accessible and therefore precludes information pertaining to the total particle. One can observe only a cross-sectional portion of the colloid, with molecular weight, M, of 882,000 ± 28,000, an electron density difference, $\Delta\rho$, of 8.1 ± 0.4 e^-/nm3, a hydration, H, of 7.92 ± 0.42 g water/g protein, and a volume, V, of (12.72 ± 0.25) x 10³/nm³. By contrast, molecular weights of whole casein micelles have been reported to range from 0.5 to 1 x 10⁹ (21). It is clear, therefore, that of the cross-sectional parameters only the electron density difference and the hydration can be directly compared with literature values. Our result for hydration of 7.92 g water/g protein is somewhat larger than the largest value reported by small-angle neutron-scattering (4.0 to 5.5) (22). Other reported values have ranged from 2 to 7, depending upon the method employed (5).

Studies by ^{2}H NMR (Table III) showed that the overall degree of tightly bound water increased nearly 2.5-fold on going from the submicellar to the micellar state. SAXS data indicate that overall hydration increases 1.4-fold, with the main increase occurring in H_L, the hydration of the loose region. The average Stokes radius of the bound water, as detected by ^{2}H NMR, increased to 4.29 nm in the micelle (Table III); in contrast, R_C from SAXS increased only slightly to 3.96 mm while R_L increased significantly. Thus, the isotropically bound water in the micelle most likely occurs outside of the core region, but well within the loose region. It should be noted that few secondary structural changes accompany the transitions that occur on micelle formation (16). Thus, gross conformational changes can be ruled out, although some small structural rearrangements may occur (16,17). All of these results indicate an increase in both bound and trapped water on going from the submicellar to the micellar state.

TABLE III. Hydration dynamics and geometry of casein bound water, isotropic mechanism [1]

Sample B Variant	Temperature °C	θ ns	nw gH$_2$O/ g protein	Radius nm
Submicelles	30	38.9	0.00652	3.64
	15	34.7	0.00824	3.05
	2	29.8	0.01201	2.55
Micelles	30	63.6	0.0165	4.29
	15	51.1	0.0225	3.48
	2	45.1	0.0282	2.93

[1]Taken from reference (27).

With regard to the electron density difference, $\Delta\rho$, this parameter remains, within error, relatively unchanged for the core region on addition of Ca^{2+}; this supports the conclusion that the internal core consists mainly of a hydrophobically rich environment. Nevertheless, the absolute electron density of this region remains, as noted above, significantly lower than in globular proteins.

The major change in electron density difference occurs for the loose region, a 30% reduction for micelles of the B variant. This is accompanied not only by an increase in hydration, H_L, but by an increase in volume, V_L. For these parameters, it appears likely that the increases are due to Ca^{2+} binding to protein electrostatic groups within this region. Moreover, the binding of Ca^{2+} occurs not only with phosphate groups but also with carboxylate groups, as shown most recently by FTIR (17). Thus, the binding of Ca^{2+} to submicelles and the subsequent transition to the micellar state does not produce more compact structures but rather more open structures. Overall, the characteristics of the submicelles appear to persist within the micelle. This conclusion is supported by recent small-angle neutron-scattering data on fresh milk micelles (29).

A note of caution is in order regarding the use of the micellar parameters, other than the hydration and electron density. As already mentioned, these do not refer to the entire particle but only to a sample portion which is restricted in size by a window of scattered intensities bounded by the lower small-angle limit of observation. They do not bear a readily defined relationship to the corresponding, but inaccessible, parameters applicable to the entire micellar particle, and therefore cannot be used to derive values for the latter. Nonetheless, the cross-sectional parameters are of value in affording an internal view of the micellar structure. The crucial comparison is between the molecular and structural parameters of the casein submicellar structure by itself (columns 2 and 3) and within the casein micelle (columns 4 and 5 of Tables I and II). Within experimental error, M_2, k, M_C and M_L are the same, V_2, V_L, R_2, and R_L increase, and V_C increases slightly; these slight changes emphasize again that the properties of the submicelles are largely preserved within the micelle.

To ascertain the spatial arrangement of the three spheres within the observed cross-sectional scattering volume observed for α_{s1}-B, the distance distribution function, p(r), was calculated from the SAXS data for casein micelles as shown in Figure 1A. Calculation of the radius of gyration from the second moment of the p(r) data (30,31) in Figure 1A, to the maximum diameter of 51.2 nm, yielded a value of 17.52 nm.

The experimental p(r) results in Figure 1A were then compared with theoretical curves calculated by the method of Glatter (30), for various geometric models. For these, the radii of the outer and inner spheres, calculated from V_2 and V_C values of column 3 of Table II, were 10.2 and 5.0 nm, respectively. The equilateral or symmetrical triangular arrangement gave the poorest fit to the experimental data (dashed line). The Cartesian coordinates for the centers of the three inhomogeneous spheres most compatible with the experimental data were found at nonsymmetrical values of (0,0), (35,0), and (18,10) (ticked line), but a better fit resulted from changing the radius of the (0,0) sphere to 12.5 nm (solid line). In fact, a radius of gyration of 17.45 nm is calculated from the theoretical p(r) curve for this inhomogeneous, irregular, triangular structure, in excellent agreement with the value of 17.52 nm found from the experimental p(r) data. It is notable that these best-fit coordinates imply interdigitation of the "loose" regions of the three submicelles (Figure 1B).

The loose, highly solvated regions of the submicelles thus retain most of their characteristics upon incorporation into the micelle. Therefore, it seems logical to assume that these loose regions will naturally form the outermost layers of the casein micelles. This means that the micellar surface would have a porous hydrophilic outer layer occurring at depths of up to 6 nm (R_L-R_C of micelles Table II). The "hairy" micelle theory (6) calls for projecting hairs of the macropeptide of κ-casein to extend outward from the micelle over average distances of 12 nm. While κ-casein most likely predominates at the surface of casein micelles, it seems unlikely that the entire loose regions, and indeed a part of the hydrophobic core, would be made up completely of the hydrophilic κ-casein macropeptide. It would seem more logical and cautious at this juncture to ascribe the non-coalescence of casein micelle, or "stability to close approach" as it was termed by Waugh (32), to a combination of steric and electrostatic forces which occur in the loose volume element of the outermost submicelles.

In fact, this feature is not unique to casein. An outer, less electron dense, region, albeit much smaller than that of the caseins, has been calculated for the compact globular protein ribonuclease (9). Indeed, changes in the surface environment of even small globular proteins are responsible for maintenance of monomeric structure as well as aggregation and denaturation. For most enzymes and structural proteins many of their associative properties are controlled by preferential solvation or preferential binding. The basic theories for these interactions were elucidated by Arakawa and Timasheff (33,34).

Variant A. As seen in Table I, the k_2 value for casein micelles of variant A was 0.167 ± 0.038 and its packing number 6.0. The cross-sectional portion of the colloid has a molecular weight, M, of 2,090,000 ± 500,000, an electron density difference, Δρ, of 9.3 ± 0.9 e⁻/nm³, a hydration, H, of 6.71 + 0.63 g water/g protein (Table I) and a volume, V, of (26.08 ± 2.39) x 10^3/nm³ (Table II). The distribution function for variant A could not be treated as was that for variant B, though it showed two separate peaks, probably a reflection of this denser packing into micelles (6:1 in place of 3:1) (Figure 2).

The molecular (Table I) and structural (Table II) parameters for A and B show many consistent differences in both submicelles and micelles. For submicelles, the various electron density differences were somewhat higher for A than for B, and this was the case also with V_2 and V_C. The various hydrations, on the other hand, were lower for A than for B in both submicelles and micelles. On going from the submicellar to the micellar state, greater differences occurred in the micelles containing A. Here, molecular weights were substantially higher for A than B, but the various radii of gyration did not increase for micelles of A as they did for those of B.

The most striking differences observed were those in the values of k. In the submicelles of A, these were more than double those of B, while in the micelles both k_2 and k_C were far

A

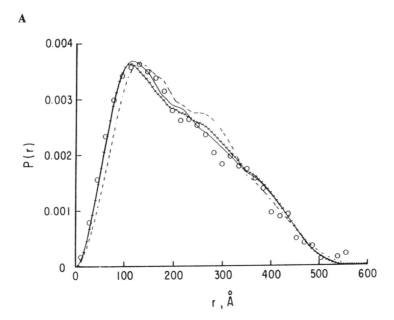

B

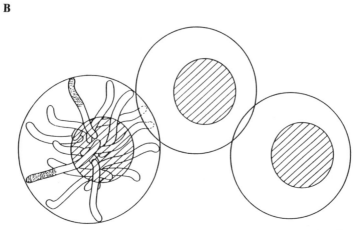

Figure 1. (A) Distance distribution of micelles. o, $p(r)$ vs r from SAXS data for micellar casein (variant B) at 16.4 mg/ml in 10 mM $CaCl_2$. Compared with theoretical curves for three inhomogeneous spheres (submicelles) all with outer radii of 10.2 nm and inner radii of 5.0 nm (━┼┼┼━) at (0, 0), (350, 0) , and (180, 100); (– – – –) in a symmetrical triangular arrangement ; (————) with two different outer radii at (0, 0; 12.5 nm), (350, 0; 10.2 nm), and (180, 0; 10.2 nm.) Theoretical curves were calculated by method of Glatter (30). (B) Schematic representation of submicelles in micellar cross section, corresponding to (————) in (A). Cross-hatched area, approximate region of higher concentration of hydrophobic side chains and higher electron density. In the lower left particle a few representative monomer chains are indicated.

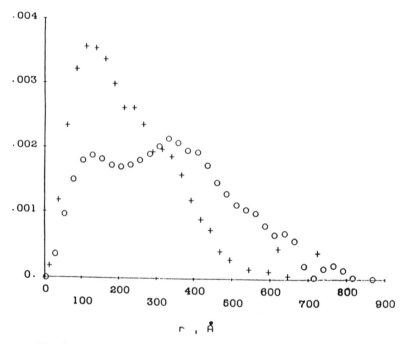

Figure 2. Distance distributions of micelles: variant A (o); variant B (+).

smaller for A than for B. These differences relate to the much higher packing number in A than in B. This larger packing number in A, as well as the larger molecular weight of the compact region and the greater elliptic eccentricity, most likely have their origins in the small but distinct differences in sequences, in calcium binding sites, and in charge distributions (35). The deletion which occurs in α_{s1}-A produces profound changes in the physical properties of the milk. The micelles are less well hydrated, and the milks have a very low heat stability (8). In addition there is a difference in overall micelle size, with micelles of A being larger than those of B (36) The high packing number disclosed by SAXS appears to be the physical basis for all of these differences in properties.

There is, however, the possibility of a completely different interpretation of the data. It could be that variant B actually does have the same packing number, 6, as A, but for reasons of interference this does not become apparent in the parameters of B. It is conceivable that the residues present in B, but absent in A, give rise to formation of a dimerlike structure, in which pairs of oppositely arranged submicelles might cause destructive interference such that half the submicelles would, in effect, remain invisible to X-rays. Only half the actual packing number (3 rather than 6) would then be observed. The differences in parameters which we found between the variants would have little part in this interpretation, but they might still be invoked to explain some of the differences in behavior of the variants. While speculative, this explanation cannot be ruled out without substantially more detailed information about casein molecular structure.

Molecular Dynamics of Casein by ^{13}C NMR. The picture presented thus far for casein micelles and submicelles is that of an open porous structure for both of these particles. The evidence supporting this feature of micelle structure comes from water binding and SAXS studies (9,27). If indeed this is the case, one would expect considerable mobility for the amino acid side chains and perhaps even for the casein backbones as well. An independent technique for studying the molecular dynamics of proteins is ^{13}C NMR.

Although the assumption of isotropic motion for protein side chains is generally an oversimplification, it may be a reasonably accurate description for immobile groups within spherical proteins and protein assemblies (such as submicelles and micelles) and can serve as a point of departure for a discussion of mobilities as detected by NMR.

Carbon-13 NMR relaxation measurements provide information concerning the mobility of chemical groups within protein molecules on the timescale of 10^{-7} to 10^{-12}s (37,38). The average rotational correlation time for backbone carbon atoms of a native protein is a reliable estimate of the correlation time for the overall tumbling of the entire protein molecule (39).

Correlation times, τ_c, that characterize the overall isotropic tumbling of casein submicelles and micelles can be estimated from the Stokes radius of the particle and the viscosity of the medium. The mere observation of high-resolution ^{13}C NMR spectra (Fig. 3) for casein submicelles and micelles strongly suggests the occurrence of considerable fast local motion within these two large particles. The decrease in mobility upon micelle formation is much less than expected from the increased protein size where, in the absence of local motion, peaks would be too broad to be detected.

T_1 values of the submicelle and micelle αCH envelopes are essentially identical (390 ms at 37°C), strongly suggesting that the dynamic state of the submicelle backbones is not affected by incorporation into casein micelles. The corresponding correlation time for an isotropic rotation model (no internal motion) is approximately 8 ns (40). Consideration of a more involved model with a 10 ps internal libration at a mean angle of 20° in addition to the overall isotropic rotational tumbling (38) results in a reduction of τ_c by a factor of 2. These values (8 or 4 ns) are comparable to those of small monomeric globular proteins, well below the 56 ns expected for the overall tumbling of the 4 nm - radius submicelles.

There appears to be considerable mobility of the protein backbones within the casein submicelle. Attempts to assess the dynamic state of casein by ^{1}H NMR (41-43) were thwarted by the extensive spectral overlap which interfered with estimation of mobility from linewidths. The broad-featured casein ^{1}H NMR spectra appeared to be the result of limited spectral resolution rather than restricted molecular mobility. Indeed, proton NMR data have suggested the existence of some fast motion within casein micelles (41-43). In one study, the

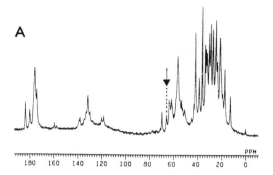

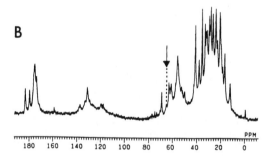

Figure 3. Natural-abundance $^{13}C\{^1H\}$ NMR spectra of submicelles (A) and micelles (B) of whole casein (variant B). The peak (indicated by the arrow) assigned to the βCH$_2$ of phosphoserine is broadened in the presence of calcium. Decreased resolution in (B) compared to (A) may result from a decrease in field homogeneity due to different magnetic susceptibilities of the micelle particle and the surrounding solution.

mobile protein regions were exclusively identified with the glycomacropeptide (GMP) segment of κ-casein, thus supporting a "hairy" micelle model, where flexible GMP chains extend into solution from the surface of rigid protein cores (41). It is unclear how such a rigid structure will at the same time be loosely packed in order to explain the ready penetration of lactose, salt and proteolytic enzymes into the micelle (1). Also, it is unlikely that the evidence of mobility in our ^{13}C NMR spectra of submicelles and micelles is solely due to the GMP which comprises less than 4% mol of whole casein and less than 0.3% protein in our samples. Other ^{1}H NMR results have indicated that only part of the observed mobility in micelles is due to κ-casein (42).

Higher mobilities are generally observed for side chain groups that are further away from the protein backbone (37-39,44). Indeed, resonances due to fast segmental motion of Lys as well as Arg and Phe (which do not occur in GMP) side chains within the casein submicelles were resolved in ^{13}C NMR spectra of both submicellar and micellar casein.

Conclusion

In summary, SAXS data for both variants argue for submicellar particles consisting of an inner, spherically symmetrical, hydrophobic, and relatively electron-dense core, surrounded by a hydrophilic and less electron-dense region, both much less dense than globular proteins, as depicted in Fig. 1A. For the casein micelles, the cross-sectional scattering volume indicates some interaction between the loose regions of adjacent submicelles. The outermost layer of the micelle is thus composed of the loose volume elements of the outer submicelles. It may be concluded that a discrete hydrophobically stabilized submicellar structure exists within the colloidal casein micelle. Such a model is supported by the high mobilities observed in ^{13}C NMR and is also in accord with the known physical and chemical properties of casein micelles including the ready diffusion of cosolutes and water.

Literature Cited

1. Schmidt, D. G. In Developments in Dairy Chemistry; Fox, P. F., Ed.; Appl. Sci., Essex, Engl,.1982; Vol. 1, p 61.
2. Davies, D. T.; Law, A. J. R. J. Dairy Res. 1980, 47, 83.
3. Farrell, H. M., Jr.; Thompson, M. P. In Calcium Binding Proteins; Thompson, M. P., Ed. CRC Press, Inc., Boca Raton, FL, 1988; Vol. 2, 117.
4. Pepper L.; Farrell, H. M., Jr. J. Dairy Sci. 1982, 65, 2259.
5. Walstra, P. J. Dairy Res. 1979, 46, 317.
6 Holt, C.; Dalgleish, D. G. J. Colloid Interface Sci. 1986, 114, 513.
7. Pessen, H.; Kumosinski, T. K.; Timasheff, S. N. Meth. Enzymol. 1973, 27, 151.
8. Thompson, M. P.; Gordon, W. G.; Boswell, R. T.; Farrell, H. M.,Jr. J. Dairy Sci. 1969, 52, 1166.
9. Kumosinski, T. F.; Pessen, H.; Farrell, H. M., Jr.; Brumberger. H. Arch. Biochem. Biophys. 1988, 266, 548.
10. Thompson, M. P. J. Dairy Sci. 1964, 47, 1261.
11. Eigel W. N.; Butler, J. E.; Ernstrom, C. A.; Farrell, H. M., Jr.; Harwalkar, V. R.; Jenness, R.; Whitney. R. McL. J. Dairy Sci. 1984, 67, 1599.
12. Lake, J. A. Acta Crystallographica 1967, 23, 191.
13. Vold, R. L.; Waugh, J. S.; Klein, M. P.; Phelps, D. E. J. Chem. Phys. 1968, 48, 3831.
14. Chou, P. Y.; Fasman, G. D. Adv. Enzymology. 1978, 47, 45.
15. Garnier, J.; Osguthorpe, D. J.; Robson B. J. Mol. Biol. 1978, 120, 97.
16. Byler, D. M.; Farrell, H. M. Jr.; Susi H. J. Dairy Sci. 1988, 71, 2622.
17. Byler, D. M.; Farrell, H. M. Jr. J. Dairy Sci. 1989, 72, 1719.
18. Luzzati, V.; Witz, J.; Nicolaieff, A. J. Mol. Biol. 1961, 3, 367.
19. Luzzati, V.; Witz, J.; Nicolaieff, A. J. Mol. Biol. 1961, 3, 379.
20. Pessen, H.; Kumosinski, T. K.; Farrell, H. M., Jr. J. Ind. Microbiol. 1988, 3, 89.
21. Schmidt, D. G.; Payens, T. A. J. In Surface and Colloid Science; Matijevic, E. Ed.; Wiley, New York, 1976; p 165.
22. Stothart, P. H.; Cebula, D. J. J. Mol. Biol. 1982, 160, 391.
23. Kuntz, I. D.; Kauzmann W. Adv. Protein Chem. 1974, 28, 239.

24. Tanford, C. In Physical Chemistry of Macromolecules; Wiley, New York, 1961;.p
     236.
25. Lumry, R.; Rosenberg A. Colloques Internationaux du Centre National de la
     Recherche Scientifique 1975, 246, 53.
26. Richards, F. M. J. Mol. Biol. 1974, 82, 1.
27. Farrell, H. M., Jr.; Pessen, H.; Kumosinski, T. K. J. Dairy Sci. 1989 72, 562.
28. Kumosinski, T. F.; Moscow, J. J.; Brown, E. M.; Farrell, H. M., Jr. Biophys. J.
     1989, 54, 333a.
29. Stothart, P. H. J. Mol. Biol. 1989, 208, 635.
30. Glatter, 0. .Acta Physica Austriaca.1980, 52, 243.
31. Pilz, I.; Glatter, 0.; Kratky, 0. Meth. Enzymol. 1979, 61, 148.
32. Waugh, D. F. In Milk Proteins; McKenzie, H. A. Ed. Academic: New York, 1971;
     Vol. 2, p 58.
33. Arakawa, T.; Timasheff, S. N. Biochemistry 1982, 21, 6536.
34. Arakawa, T.; Timasheff, S. N. Biochemistry 1984, 23, 5912.
35. Farrell, H. M. Jr; Kumosinski, T. K.; Pulaski, P.; Thompson, M. P. Archives
     Biochem. Biophys. 1988, 265, 146.
36. Dewan, R. K.; Chudgar, A.; Bloomfield, V. A.; Morr, C. V. J. Dairy Sci. 1974, 57,
     394.
37. Wüthrich, K. NMR in Biological Research: Peptides and Proteins; North-Holland:
     Amsterdam, 1976.
38. Howarth, 0. W.; Lilley, D. M. J. Prog. NMR Spectrosc. 1978, 12, 1.
39. Richarz, R.; Nagayama, K.; Wüthrich, K. Biochemistry 1980, 19, 5189.
40. Doddrell, D.; Glushko, V.; Allerhand, A. J. Chem. Phys. 1972, 56, 3683.
41. Griffin, M. C. A.; Roberts, G. C. K. Biochem. J. 1985, 228, 273.
42. Rollema, H. S.; Brinkhuis, J. A.; Vreeman, H. J. Neth. Milk Dairy J. 1988, 42, 233.
43. Rollema, H. S.; Brinkhuis, J. A. J. Dairy Res. 1989, 56, 417.
44. Allerhand, A. Meth. Enzymol. 1979, 61, 458.

RECEIVED August 14, 1990

Chapter 14

Genetic Engineering of Bovine κ-Casein To Enhance Proteolysis by Chymosin

Sangsuk Oh and Tom Richardson

Department of Food Science and Technology, University of California,
Davis, CA 95616

k-Casein cDNA was mutated to change the chymosin
sensitive   site from a Phe(105)-Met(106) bond to a
Phe(105)-Phe(106) bond which, in theory, should be
attacked at a faster rate by acid proteases. Mutant k-
casein and normal k-casein were expressed in E.coli
strain AR68 using the secretion vector, pIN-III-ompA.
The expressed k-caseins were extracted in urea buffer
and partially purified using DE 52 anion exchange
chromatography. Partially purified k-caseins were
hydrolyzed with chymosin at 30°C. Initial hydrolysis
rates were compared. The mutant k-casein (Phe(105)-
Phe(106)) was hydrolyzed approximately 80 percent
faster than the wild-type (Phe(105)-Met(106)) k-casein
as determined using western blots, followed by
immunochemical staining and laser gel scanning.

The stabilities of proteins to enzymatic attack are to a large extent
dictated by the structure of the amino acids on either side of the
susceptible bond and, in some cases, on the primary sequence around
the scissile linkage. Moreover, exposure of the bond to enzymatic
attack is dependent upon its steric availability. Thus, the three-
dimensional structure of a protein may be important for enzymatic
attack.
    The rate of hydrolysis of a given bond can have wide-ranging
ramifications on the functional characteristics of food proteins and
on the quality of resultant food products. For example, in the
coagulation of milk to make cheese, the Phe(105)-Met(106) bond of k-
casein is attacked by chymosin (rennin). This results in the loss of
a polar peptide from the C-terminus of k-casein which destabilizes
the casein micelles comprised of α_{s1}-, α_{s2}-, β- and k-caseins (1).
In the presence of calcium ions, the destabilized casein micelles
coagulate to form cheese curd. Chymosin has a relatively high
initial specificity for the Phe(105)-Met(106) bond of k-casein,
having a high ratio of activity towards this bond compared to general

0097–6156/91/0454–0195$06.00/0
© 1991 American Chemical Society

proteolysis. Proteolytic enzymes with more general proteolytic
activity, such as trypsin, tend to release bitter peptides from the
caseins (2), thus having a detrimental effect on cheese flavor.
Subsequent aging of the cheese is thought to be due in part to
controlled proteolysis of the caseins by clotting and microbial
proteases. Basically good cheese is in part a function of the
susceptibilities of various bonds in the caseins to hydrolysis and
the rates of their cleavage (1,2). The specific action of chymosin
on k-casein is also important because general proteolysis of chymosin
would result in increased soluble peptides in the whey, resulting in
economic loss.

With the use of genetic engineering techniques, it may be possible
to alter the rate at which a susceptible bond is attacked and thereby
alter the rate at which cheese flavor and texture develops or cheese
ages. The substitution of about 20% of an endogenous casein with a
casein mutant may have a significant effect on the functionality of
the resultant casein mix (3). Thus it may be possible to genetically
engineer milk to alter many fundamental dairy processes including the
aging of cheese.

Hydrolysis of k-Casein by Chymosin. The amino acids between residues
97 and 112 are well conserved in bovine, ovine and caprine k-casein
thereby demonstrating their functional importance (4, 5). The
specificity of chymosin for k-casein is related in part to the
presence of the cluster of cationic residues at position 97-112 in k-
casein which is comprised of one arginine, three histidines, and two
lysine residues (6, 7). The dipeptide, Phe-Met, is not hydrolyzed by
chymosin (8), nor indeed are tri- and tetra-peptides even though they
contain the Phe-Met bond (9). A pentapeptide sequence found in k-
casein, Ser-Leu-Phe-Met-Ala-OMe, however, was hydrolyzed more rapidly
by chymosin than the tetra peptide Leu-Phe-Met-Ala-OMe at the Phe-Met
bond (9, 10). This shows not only the length but the composition and
sequence of the peptide substrate are important in determining the
chymosin and k-casein interaction (Figure 1).

Based on the results derived from the studies of the subsite
specificity of pepsin and of the k-casein and chymosin interactions
using various sized synthetic peptides, We propose that the Phe-Phe
bond should be a better cleavage site than Phe(105)-Met(106) for
chymosin in the context of k-casein. If it is not better, then there
may be factors involved in k-casein site specificity that overrides
this presumed preference for the Phe-Phe bond.

The purpose of this research was to mutate the Phe(105)-Met(106)
bond of k-casein to a Phe(105)-Phe(106) bond and to study the effect
of this alteration on the hydrolysis of the k-casein mutant, compared
to a control, by chymosin.

Materials and Methods

Bacterial Strains and Plasmids. The E.coli strain JM107 was
purchased from Amersham Corp., Arlington Heights, Il. and the E.coli
strain AR68 was a generous gift from Dr. Shatzman, Smith Kline and
French Laboratories. Plasmids M13mp18 RF(replicative form) and
M13mp19 RF were obtained from Boeringer Mannheim, Indianapolis, In.
and the pIN-III-ompA1 was obtained from Dr. Inouye, N.J. Medical
College, Newark, N.J..

<u>Single Stranded Template Preparation.</u> Bovine cDNA for k-casein has
been cloned and expressed (11). The plasmid, pKR76, contains a full
length cDNA fragment which codes for mature bovine k-casein. The
pKR76 was digested with <u>Msp</u>I endonuclease. The resultant 716 base
pair (bp) DNA fragment from an <u>Msp</u>I digest of pKR76 was purified
after electrophoresis in low temperature agarose gel and by elution
from the agarose using an Elutip-d mini column (Schleicher and
Schuell, Keene, NH). The DNA fragment was subsequently digested with
<u>Nla</u>IV endonuclease, and the resultant blunt-sticky ended fragment was
purified. The purified <u>Msp</u>I-<u>Nla</u>IV fragment was inserted between <u>Sma</u>I
and <u>Acc</u>I sites of bacteriophage M13mp18 RF. The ligation,
transformation, and screening of recombinant M13 bacteriophage
plaques were performed according to the protocol described by Messing
(12). From the selected phage plaques, single stranded template was
prepared (12).

<u>Mutagenesis Reaction.</u> An oligonucleotide (24mer) which induces the
change of the Phe(105)-Met(106) bond to a Phe(105)-Phe(106) in k-
casein was prepared by J. Pressley at the Protein Structure Research
Laboratories, University of California at Davis (Figure 2). The
gapped duplex method (Figure 3) was used for mutagenesis (13). Five
hundred nanograms of M13 mp18 RF were double digested with <u>Hind</u>III
and <u>Eco</u>RI endonucleases and purified using an Elutip-d column. Ten
nanograms of phosphorylated oligonucleotide, 4µl ligase buffer, and
100 ng of single stranded template with the k-casein cDNA insert were
added to the mix and H_2O was added with mixing to 40 µl. This
solution was heated to $95^{\circ}C$ for 10 min, cooled to room temperature
for 30 min and then kept on ice for 10 min. To this solution were
added 1µl of 2.5 mM dNTPs, 1 unit of T4 DNA ligase, 1 unit of Klenow
fragment, 1µl of ligase buffer, 2µl of 10 mM ATP and 4µl H_2O and the
solution was incubated at $4^{\circ}C$ for 30 min, at $12^{\circ}C$ for 1 h, and at
$37^{\circ}C$ for 30 min.
      Ten microliters of above solution were mixed with 0.3 ml of $CaCl_2$-
treated competent <u>E.coli</u> cells, JM 107, and the culture was kept on
ice for 40 min. This mixture was heat-shocked at $42^{\circ}C$ for 3 min.
Two tenths milliliters of exponentially growing fresh JM 107 cells
and 10 ml of 2 x YT media were mixed together with the heat-shocked
competent cells in a 50 ml flask. This mixture was incubated at $37^{\circ}C$
overnight. The cell culture was transfered to Eppendorf tubes and
centrifuged for 5 min in a microfuge. A series of dilutions (10^{1-}
10^{10}) of the cell culture supernatant liquid were prepared. The
diluted supernatant portions were added to 3.5 ml of H top agar
containing 300µl of fresh JM 107 cells, 10 µl of 200mM IPTG
(isopropyl-β-D-thiogalctopyrano-side), and 50µl of X-gal (5-bromo-4-
chloro-3-indolyl-β-D-galactopyranoside) (20mg/ml in dimethyl
formamide). This mix was gently shaken and poured onto LB plates(10
g tryptone, 5 g yeast extract, 5 g NaCl, 1ml 1N NaOH, and 15 g agar
per liter). The LB plates were kept at room temperature for 15 min
and incubated at $37^{\circ}C$ for 8 h.

<u>Screening of Mutants.</u> Zoller and Smith's method (14) was used for
plaque hybridization. Plates which contained 100 - 400 plaques were
chosen for plaque lifts. The moist nitrocellulose filters
(Schleicher & Schuell, Keene, New Hampshire) were placed on the LB

```
His-Pro-His-Pro-His-Leu-Ser|-Phe-Met-|Ala-Ile-Pro-Pro-Lys-Lys      Kcat/Km
(98)                        |Phe-Met  |              (112)          0.00
                        Ser|-Phe-Met-|Ala-Ile                       0.04
                        Ser|-Phe-Met-|Ala-Ile-Pro                   0.11
                    Leu-Ser|-Phe-Met-|Ala-Ile                      21.6
                His-Leu-Ser|-Phe-Met-|Ala-Ile                      31
                    Leu-Ser|-Phe-Met-|Ala-Ile-Pro-Pro              100
            Pro-His-Leu-Ser|-Phe-Met-|Ala-Ile                     100
His-Pro-His-Pro-His-Leu-Ser|-Phe-Met-|Ala-Ile-Pro-Pro-Lys-Lys   2500
```

Figure 1. The relative rates of hydrolysis of some peptides
sequences found in k-casein by chymosin (18,19,20).

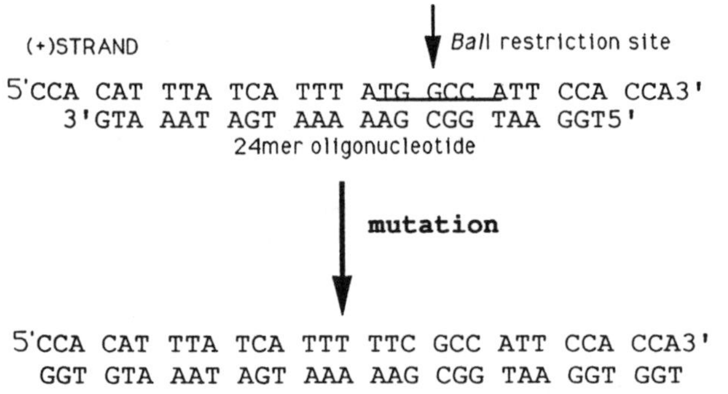

(+)STRAND   ↓ *Ball* restriction site

5'CCA CAT TTA TCA TTT ATG GCC ATT CCA CCA3'
3'GTA AAT AGT AAA AAG CGG TAA GGT5'
                24mer oligonucleotide

                    ↓ mutation

5'CCA CAT TTA TCA TTT TTC GCC ATT CCA CCA3'
GGT GTA AAT AGT AAA AAG CGG TAA GGT GGT

Figure 2. Oligonucleotide sequence to induce the change of
Met(106)(ATG) to Phe(106)(TTC) in k-casein

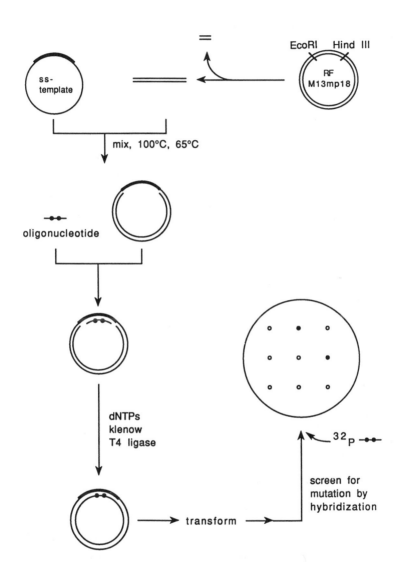

Figure 3. General scheme of oligonucleotide-directed mutagenesis
for the Phe(105)-Phe(106) mutation. Single stranded DNA with k-
casein cDNA insert was mixed with linearized M13mp18 RF treated
with HindIII and EcoRI. After the mixture was boiled and cooled
to 65°C, oligonucleotide was added and the mixture was cooled to
room temperature. The gapped heteroduplex DNA was filled-in and
ligated using Klenow fragment and T4 DNA ligase. The mixture was
then transformed into competent E.coli cells. Radiolabeled
oligonucleotide was used to screen for mutant k-casein.

plates for 1 min to obtain the first replica, and 3 min for the
second replica. These nitrocellulose filters were then baked in an
80°C vacuum oven for 2h. The nitrocellulose filters were placed in
prehybridization solution (6x SSC, 10 x Denhardts buffer, 0.2 %
SDS)· The nitrocellulose filters in the prehybridization solution
were incubated at 65°C for 1h.

The nitrocellulose filters were placed subsequently in
hybridization solution (6 x SSC, 10 x Denhardts buffer, ^{32}P labeled
oligonucleotide) and incubated at room temperature for 1h. The
hybridization solution was dicarded and the nitrocellulose filters
were washed with 6 x SSC for a total of 15 min at room temperature.
Mismatched primers were removed from the filters using 45°C and 55°C
washing in 6 x SSC for 10 min. The washing temperature was
determined according to Suggs et al.(15). The nitrocellulose filters
were dried slightly and exposed overnight to X-ray film (Kodak X-Omat
AR) with the aid of an intensifying screen (DuPont, Lightning)
overnight. The k-casein cDNA from the putative mutant was subcloned
into M13 mp18 bacteriophage and sequenced using the dideoxy chain
termination method (16).

The following conditions were used for labeling the
oligonucleotide used for screening. Two hundred nanograms of
oligonucleotide, 10µl of 50mM MgCl$_2$, 5µl of 1 M Tris-HCl pH 7.6, 5 µl
of 200 mM 2-mercaptoethanol, 100 µCi [gamma-^{32}P]ATP, and 10 units of
T4 polynucleotide kinase in a final volume of 50 µl were incubated at
37°C for 1 h. The reaction was stopped by heating the mixture at
65°C for 10 min.

Construction of Expression Vector, pKOR. Kappa casein cDNA was
inserted into the RF of M13mp18. This recombinant RF was purified
using a CsCl-EtBr gradient (17). The expression vector pIN-III-ompA$_1$
(18) was used to obtain expression of k-casein and its mutant in
E.coli strain AR68 which is protease deficient.

The expression vector, pIN-III-ompA1 was purified using a CsCl-
EtBr gradient. Five micrograms of pIN-III-ompA1 in 100 µl buffer was
completely cleaved with EcoRI endonuclease. The restriction enzyme
was inactivated by heating 10 min at 65°C. The linearized DNA
segment was isolated using an Elutip-d column after low temperature
agarose gel electrophoresis. The linearized DNA was again cleaved
using HindIII endonulease and purified using an Elutip-d column. The
k-casein cDNA insert of pKC was isolated after digestion of the
plasmid with EcoRI and HindIII endonucleases. Five micrograms of pKC
was cleaved with EcoRI, and the linearized DNA was precipitated using
ethanol. The linearized DNA redissolved in water was cleaved with
HindIII endonuclease. The k-casein insert cDNA was isolated using an
Elutip-d column after low-temperature agarose gel electrophoresis and
dissolved in 100µl H$_2$O. After ligation, plasmid DNA was transformed
into competent E.coli JM 107 and AR 68 cells. Selection of the
transformants was accomplished using the ampicillin resistant marker
of the expression vector. From the individual transformants, the
plasmid was purified using the miniscreen procedure. Purified DNA
was digested with BalI endonuclease and subjected to agarose gel
electrophoresis to verify the insertion of k-casein cDNA.

Detection of k-Casein Expressed in E.coli. Transformant cells
harboring pKOR-phe were selected and grown in 1 ml LB-ampicillin
broth to detect the expression of k-casein in E.coli by SDS-PAGE,
western blot, and immunochemical staining (19).
   Cell pellets from 1 ml of the overnight culture derived from
colonies containing expression plasmids were collected and
resuspended each in 100µl of SDS sample buffer. The mixtures were
boiled for 5 min and 50µl aliquots for each sample were subjected to
electrophoresis in 10% SDS-PAGE. The proteins were blotted on
nitrocellulose membranes and tested for the presence of k-casein as
described by Hawkes (19) with antibodies against bovine k-casein.

Partial Purification of k-Casein. The E.coli strain harboring
pKOR1(wild type) or pKOR-phe was grown in LB medium containing
ampicillin (50 µg/ml) to an O.D.$_{600}$ of 0.25. One liter cultures,
induced after addition of IPTG (1 mM), were grown 4 h. Cells were
harvested after centrifugation (5,000 x g). Cell pellets were
suspended in 20 ml of 30 mM Tris-HCl (pH8.0) containing 5 mM EDTA,
250 µg lysozyme per milliliter, and phenylmethylsulfonylfluoride
(PMSF) to 1 mM. The cell suspension was rapidly frozen in dry ice-
methanol and thawed at 37 °C in a water bath three times. The
resulting lysed cell paste was homogenized and extracted twice with
the same volume of 8M urea containing 30 mM Tris-HCl (pH 8.0), 5 mM
EDTA, and 150 mM NaCl. The 8M urea extract was applied to a DE52
(Whatman, U.K.) ion exchange column.
   The DE 52 anion exchange column was prepared using a Bio-Rad Econo
column (20 cm, 15 mm ID). The column was equilibrated with running
buffer (3M urea, 30 mM Tris, pH 8.0, 5mM EDTA, 1 mM DTT). Fractions
were eluted using a 0 to 1 M NaCl linear gradient. Thirty drop
fractions were collected and analyzed using dot blot and
immunochemical methods. Fractions giving positive signals were
further analyzed using SDS-PAGE, western blot and immunochemical
staining.
   Partially purified k-casein, which was expressed in E.coli AR 68
harboring pKOR1 or pKOR-phe, was used for digestion with chymosin
(Sigma, St. Louis, MO). After DE 52 ion exchange chromatography of
the k-casein extract, fractions that were positive for k-casein
antibody were pooled and dialyzed in a dialysis bag (M.W. cut-off
12,000 - 14,000) against phosphate buffer (pH 6.8) at 4°C overnight.
Ice-cold 100% TCA was added slowly to a final concentration of 5%.
The resultant pellets were dried in vacuo after washing with 70%
ethanol. The pellets were dissolved in the same buffer containing 3M
urea and redialysed against phosphate buffer (pH 6.8). Twenty
microliters of these samples were incubated with 0.1 unit of chymosin
at 30°C for 0, 1, 2, 5, or 10 min. Twenty microliters of SDS sample
buffer was added and the mixture boiled for 2 min. Samples were
loaded onto 10% SDS polyacrylamide gels. After the western blot, the
nitrocellulose membranes were analyzed using the immunochemical
method employed to detect k-casein as well as mutant k-casein (Phe-
Phe bond). The same concentrations of k-caseins were used.
Intensities of k-casein bands on the nitrocellulose were compared
using a laser densitometer (LKB, Piscataway, NJ). Initial rates of
hydrolysis of the wild type and mutant k-casein by chymosin were
compared.

Results and Discussion

Template Preparation. The MspI fragment of pKR76 (11) which contains whole mature k-casein cDNA was further cleaved with NlaIV endonuclease. This MspI and NlaIV endonuclease cleaved fragment which has a blunt end at one end and a sticky end at the other end was purified using low temperature agarose gel electrophoresis coupled with the use of an Elutip-d minicolumn (Schleicher & Schuell Inc., Keene, NH).

The MspI - NlaIV cleaved fragment of pKR76 was ligated into the SmaI - AccI cleaved RF of M13mp18 bacteriophage. After ligation, the mixture was used to transform competent cells of E.coli JM107. The RF of the recombinant bacteriophage was purified. The purified RF was analysed using restriction enzyme patterns.

The recombinant plasmid pKC contains a k-casein cDNA insert. EcoRI - HindIII cleaved pKC revealed a 660 bp band after gel electophoresis on 1.2 % agarose. The presence of k-casein cDNA insert was also confirmed by PstI digestion followed by gel electrophoresis in 1.2% agarose. Ten RF preparations out of ten white plaques showed that they all have the insert.

Recombinant RF was transformed into competent E.coli JM 107. Single stranded recombinant phage DNA was prepared from recombinant RF phage supernatant fluid (12).

Synthesis and Purification of Oligonucleotide. The oligonucleotide (24mer) sequence which induced the change of a Phe(105)-Met(106) bond to a Phe(105)-Phe(106) bond was based on the (+) strand of the k-casein cDNA insert sequence. The oligonucleotide was designed so that the mismatch is located near the middle of the molecule, because placement of the mismatch in the middle yields the greatest binding differential between a perfectly matched duplex and a mismatched duplex. With these considerations in mind, an oligonucleotide was synthesized (Figure 2). The oligonucleotide mixture was purified using electrophoresis in 20% polyacrylamide gel (19:1, acrylamide:bisacrylamide) containing 7M urea and 1 x TBE buffer (90 mM Tris, 65 mM boric acid, 2.5 mM EDTA, pH 8.3).

The slowest moving band was excised from the gel. The gel fragment was diced and the DNA was eluted by diffusion at 37°C overnight into Maxam-Gilbert gel elution buffer (0.5 M ammonium acetate, 0.01 M magnesium acetate, 0.1 % (W/V) SDS and 0.1 mM EDTA) (20). Desalting of the extract was carried out with a Sep-Pak C_{18} mini column (Waters Inc., Milford, MA) (21). The Sep-Pak mini column was equilibrated with 5 ml of TE buffer (pH7.9). The sample was loaded and the Sep-pak minicolumn was washed with 5 ml of TE buffer. The oligomer was eluted with 40 % acetonitrile in water. The first 0.5 ml was discarded and the next 1.5 ml of eluant was collected. The oligomer solution was dried under vacuum.

Oligonucleotide-directed Mutagenesis. The RF DNA of M13mp18 bacteriophage was cleaved with HindIII and EcoRI endonucleases. Single stranded template DNA was mixed with the linearized, double cleaved RF DNA of M13mp18 bacteriophage, resulting in the gapped heteroduplex (Figure 3). The mutagenic oligonucleotide was annealed onto the single-stranded DNA of the gapped duplex and extended by DNA polymerase 'large fragment'. After Transfromation into competent cells of JM 107, the culture was plated onto an LB plate.

Screening of mutants was performed using plaque hybridization. After hybridization, the membrane was washed at 55°C for 10 min, then autoradiographed. Twenty positive plaques were revealed on the autoradiogram. The efficiency of mutagenesis (ATG to TTT; two point mutation) is 16 %. We decided to use the 'gapped heteroduplex' method to increase the circular closed double stranded DNA, because the 'gapped heteroduplex' method was easily accessible at this point of the research. Many other site-directed mutagenesis methods in which the mutagenesis efficiency has been increased up to 90 % (22) have since been developed.

Another way to screen for mutant DNA is via restriction endonuclease digestions. For the Phe-Phe mutation, the oligonucleotide was designed to change CAT TTA TCA TTT ATG GCC A TT CCA to CAT TTA TCA TTT TTC GCC ATT CCA. BalI endonuclease cleaves TGGCCA. With successful mutagenesis, TGGCCA is changed to TCGCCA which is no longer a Bal I endonuclease cleavage site. Double stranded DNA from two putative mutant plaques was isolated and rescreened for the mutation by the restriction digestion with BalI.

As the last step to confirm the change of ATG GCC to TTC GCC, sequencing was performed using the dideoxy termination method. Double stranded mutant DNA was prepared from one mutant phage and digested with PstI and EcoRI endonucleases. The 450 bp and 230 bp fragments were subcloned into M13mp18 RF. Universal primer (Pharmacia, Piscataway, NJ), and the synthesized nucleotide was used as primer for sequencing. The change of the Phe(105)-Met(106) bond to Phe(105)-Phe(106) is shown in Figure 4. There is no mutation in any other region of the k-casein cDNA.

Construction of pKOR1 and pKOR-phe. The EcoRI and HindIII double digested fragments of pKC encodes for seven additional aminoacids at the amino terminus of the k-casein. The resulting recombinant plasmid, pKOR, would have all the leader sequence of the omp gene (63bp), the first amino acid sequence of the mature outer membrane protein (omp) (3 bp), a part of the multicloning sites of m13mp18 bacteriophage (18 bp), the last amino acid of the bovine k-casein signal peptide (3bp) and the mature bovine k-casein sequence (504 bp) (Figure 5). After translocation of this recombinant gene, there would be two possible routes for the gene products. If signal peptide processing occurs at the expected position as shown in Figure 5, the expressed k-casein would have eight additional amino acids at the amino terminus. If not processed as proposed, there would be 29 additional amino acids at the amino terminus.

The expression vector, pIN-III-ompA1, is derived from pBR322, a plasmid produced in roughly 30 copies per bacterial cell (23). It employs the efficient lpp (lipophosphoprotein) gene promoter (24) and the lac UV5 promotor operator (lac po) downstream of lpp. Downstream of this tandem promoter region there is a multiple restriction site linker. Cloning the k-casein gene with a compatible reading frame, therefore, resulted in the usage of the lipoprotein Shine-Dalgano sequence, the initiation codon in addition to a termination codon and a rho-independent efficient transcription termination signal.

The HindIII and EcoRI endonuclease digested k-casein cDNA fragment was force-cloned into the HindIII-EcoRI endonuclease cleaved

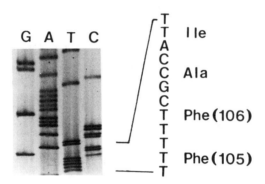

Figure 4. Portion of relevant DNA sequence of mutant pKC-phe.
The change of ATG to TTC (106) is shown at the bottom of the
pattern.

Figure 5. DNA sequence of ompA signal peptide, linker and k-casein
cDNA in the 5' region. The thick solid arrow denotes the signal
peptide cleavage site. There are 21 amino acids from the ompA
signal peptide, 6 amino acids from the linkers, 1 of k-casein
signal peptide) and the mature k-casein.

expression vector, pIN-III-ompA1 (Figure 6). Thus, k-casein cDNA fragments were ligated to the expression vector, pIN-III-ompA1 which enables the expression of k-casein under the control of the vector. The recombinant plasmid, pKOR1 and pKOR-phe, were confirmed by restriction mapping.

Quantitation of Expressed k-Casein. Expressed k-casein in E.coli AR 68 harboring pKOR1 was quantitated using laser gel scanning after SDS-PAGE, western blot and immunochemical staining. A series of increasing amounts of standard k-casein were used to prepare a standard curve. Cell extracts from 1 ml of culture after induction with IPTG was loaded onto SDS polyacrylamide gels. From the standard curve, the expressed k-casein was estimated to be 2 mg/l of medium. This amount of expressed k-casein is clearly less than that of other proteins expressed using the same system. Two reasons may explain the lower than expected expression level. First, k-casein is amphiphilic with an N-terminal hydrophobic domain and a C-terminal polar domain. Thus, it may act like a detergent and be toxic to the host. Second, k-casein is not a globular protein. The degree of randomness in its structure may yield a protein that is very susceptible to proteolytic activity. Consequently, random proteolysis of expressed k-casein in host cells might occur.

Partial Purification and Hydrolysis of expressed k-Casein by Chymosin. For the purification of k-casein, a 4-h induction period with IPTG was used. All proteins in E.coli AR68 including expressed k-casein were extracted with 6 M urea buffer, which solubilized more than 90% of the total k-casein expressed. Fractions were analyzed using western blot and immunochemical staining utilizing a polyclonal bovine casein antibody after anion exchange chromatography. Fractions from 17 to 21 elicited strong positive signals (Figure 7). Greater than 90% of the host proteins were removed based on a calculation using UV absorbance at 280 nm for protein concentration.

Mixtures of peptides can be separated by electrophoresis. If the procedure is carefully controlled, the pattern obtained is characteristic of the particular protein from which the peptides were derived (25). When bovine k-casein is digested with chymosin, para k-casein and a macropeptide are the resultant products.

Appropriate positive fractions from DE52 anion chromatography were pooled and concentrated using TCA (trichloroacetic acid) precipitation. A series of samples, wild type (Phe(105)-Met(106)) or mutant (Phe(105)-Phe(106)) k-casein, were hydrolyzed with 0.1 unit of chymosin. After hydrolysis, each sample was analyzed using SDS-PAGE, western blot and immunochemical staining. From Figure 8, it is clear that the mutant k-casein (Phe(105)-Phe(106)) is hydrolyzed faster than the wild type k-casein (Phe(105)-Met(106)). Each k-casein band was analyzed further using laser densitometry and plotted against hydrolysis time. From the initial part of the plots, it was possible to calculate the initial hydrolysis rates of wild type and mutant k-caseins. The average initial hydrolysis rate of mutant k-casein (Phe(105)-Phe(106)) in duplicate experiments was about 1.80X faster than for the wild type k-casein (Phe(105)-Met(106)) (Figure 9). From Figure 9, it is evident that the concentration of the k-casein in the reaction mixture was on the order of 10-6M. The K_m of chymosin for k-casein ranges between 0.67-5.4 x 10^{-5}M (26). Consequently, the k-

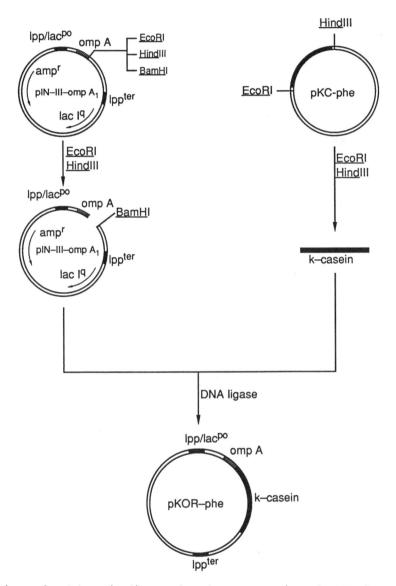

Figure 6.   Schematic diagram for the construction of pKOR-phe.
Both k-casein cDNA and linearized pIN-III-ompA₁ were purified by
double digestion with EcoRI and HindIII endonucleases.   Two
fragments were ligated using T4 DNA ligase resulting in a plasmid
for expression of mutant k-casein in E.coli, pKOR-phe.

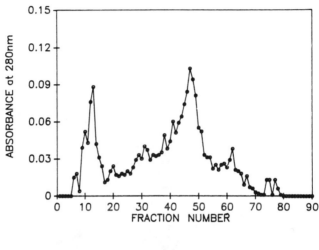

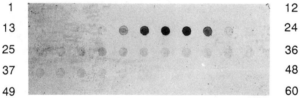

DOT BLOT OF DE52 FRACTIONS

Figure 7. DE 52 anion exchange chromatography and dot blot analyses of fractions. k-Casein expressed in E.coli AR68 was extracted with 6 M urea buffer (pH 8.0). The urea extract was applied to DE52 anion exchange chromatography and eluted using 0-1M NaCl linear gradient. Fractions were analyzed using dot blot and immunochemical staining. Positive fractions are shown at the bottom of the figure.

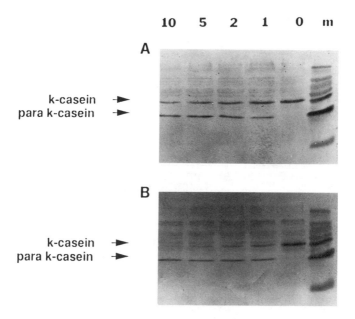

Figure 8.   Comparison of chymosin hydrolysis pattern for mutant
(Phe(105)-Phe(106)) and wild type (Phe(105)-Met(106)) k-caseins.
Partially purified wild type and mutant k-caseins were hydrolysed
with 0.1 unit of chymosin for 0, 1, 2, 5 and 10 min at 30°C.   Each
sample was analyzed by SDS-PAGE, western blot and immunochemical
staining.   A, various intervals of hydrolysis of wild type
k-casein; B, various intervals of hydrolysis of mutant k-casein.

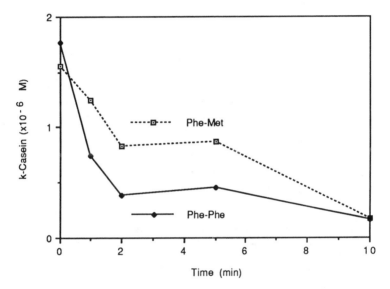

Figure 9. Hydrolysis rates of wild type (Phe(105)-Met(106)) and mutant (Phe(105)-Phe(106)) k-casein. Partially purified wild type (closed squares) and mutant (closed circles) k-caseins were hydrolyzed with chymosin at 30°C. The concentrations of hydrolyzed k-casein were plotted based on calculations from intensities of the k-casein and para-k-casein bands using laser densitometry.

casein in the reaction mixture would be expected to be near the
first-order region of the chymosin activity. However, first order
plots of the hydrolysis data were curvilinear indicating complex
kinetics. Therefore, initial rates of hydrolyses were used to
compare activities on wild type and mutant proteins.
     A number of studies have been published in which enzymes have been
engineered to modify their activities (27). However, this is one of
the first times that a protein substrate has been engineered to
enhance its susceptibility to proteolysis.
     k-Casein is on the surface of the casein micelle. It protects the
casein micelle from aggregating. The C-terminal part of k-casein is
very hydrophilic, particularly when glycosylated. This C-terminal
part of the molecule may project from the surface of the micelles and
behave as a random-coil polymer chain. Consequently, colloidal
stability of casein micelles is maintained by steric and charge
repulsion at the micellar surface. When casein micelles are treated
with chymosin the C-terminal moiety of k-casein is released into the
medium resulting in aggregates of casein micelles. It is not
difficult to imagine that casein micelles containing mutant k-caseins
on their surface would be destabilized faster than those with normal
k-casein on their surface when treated with chymosin. Therefore,
less chymosin might be used in the cheese-making process. The
average hydrophobicity of k-casein change little by mutation of
methionine to phenylalanine. Thus the physical properties of the
resultant para-k-caseins should vary to a slight degree.

Conclusions
     k-Casein cDNA was manipulated to enhance the enzyme specificity as
a substrate for chymosin. Mutagenesis was performed using the
'gapped duplex' method (Kramer, 1984). The mutation was to change
the chymosin sensitive site from a Phe(105)-Met(106) to Phe(105)-
Phe(106) which, in theory, should be attacked at a faster rate by
acid proteases. DNA sequencing confirmed that mutagenesis was
successful and that there was no mutation in any other region of the
k-casein cDNA.
     Mutant and wild type k-casein cDNAs were inserted into the
expression vector, pIN-III-ompA$_1$, to express k-casein in E.coli AR68,
which is a protease deficient strain. An expression level of 2 mg/l
of medium was obtained.
     Partially purified k-caseins of mutant and wild type were analyzed
after hydrolysis with chymosin. The data suggest that the mutant k-
casein (Phe(105)-Phe(106)) is hydrolyzed 1.80X faster than wild type
(Phe(105)-Met(106)) when the initial hydrolysis rates were compared.
This study has shown that mutant k-casein containing the Phe(105)-
Phe(106) bond of k-casein is a better substrate for chymosin than
wild-type k-casein.

Acknowledgments
     This research was supported by the National Dairy Promotion and
Research Board, the Peter J. Shield's Endowment Fund and the College
of Agricultural and Environmental Sciences, University of California,
Davis 95616.

Literature Cited
1. Dalgleish, D.G. In <u>Developments in Dairy Chemistry</u>, Vol. 1; Fox, P.F., Ed; Applied Sci.; New York, 1982; p 157.
2. Law, B.A. In <u>Advances in the Microbiology and Biochemistry of Cheese and Fermented Milk</u>; Daries, F.L. and Law, E.A Eds.; Applied Sci.; New York, 1984; p 187, 209.
3. Jimenez-Flores, R.J.; Richardson, T. <u>J. Dairy Sci.</u> 1988, <u>21</u>, 2640.
4. Jolles, P.; Henschen, A. <u>Trends Bioch. Sci.</u>, 1982, <u>7</u>, 325.
5. Brignon, G.; Chtourou, A.; Ribadeau-Dumas, R. <u>FEBS Lett.</u>, 1985, <u>188</u>, 48.
6. Visser, S.; van Rooijen, P.J.; Schettenkerk, C.; Kerling, K.E.T. <u>Biochem. Biophys. acta.</u>, 1976, <u>438</u>, 265.
7. Visser, S.; van Rooijen, P.J.; Slangen, K.J. <u>Eur. J. Bioch.</u>, 1980, <u>108</u>, 415.
8. Vonick, I.M.; Fruton, J.S. <u>Proc. Nat. Acad. Sci. USA</u>, 1971, <u>68</u>, 257.
9. Hill, R.D. <u>J. Dairy Sci.</u>, 1969, <u>36</u>, 409.
10. Hill, R.D. <u>Bioch. Biophys. Res. Comm.</u>, 1968, <u>33</u>, 659.
11. Kang, Y.S.; Richardson, T. <u>J. Dairy Sci.</u>, 1988, <u>71</u>, 29.
12. Messing, J. <u>Methods Enzymol.</u>, 1983, <u>101</u>, 20.
13. Kramer, W.; Drustra, V.; Jansen, H.-W.; Kramer, B.; Pflugfelder, M.; Fritz, H.-J. <u>Nucleic Acids Res.</u>, 1984, <u>13</u>, 4431.
14. Zoller, M.J.; Smith, M. <u>DNA</u>, 1984, <u>3</u>, 479.
15. Suggs, S.; Wallace, B.; Hirose, T.M.; Kawashima, E.; Itakura, K. <u>Proc. Natl. Acad. Sci.   USA.</u>, 1981, <u>78</u>, 6613.
16. Sanger, F.; Nicklen, S.; Coulson, A.R. <u>Proc. Natl. Acad. Sci. USA.</u>, 1977, <u>74</u>, 5463.
17. Maniatis, T.; Fritsch, E.; Sandbrook, J. In <u>Molecular Cloning</u>; Cold Spring Harbor Laboratory: New York; 1982.
18. Ghrayeb, J.; Kimura, H.; Takahara, H.; Hsiuiry, H.; Masui, Y.; Inouye, M. <u>EMBO J.</u>, 1984, <u>3</u>, 2473.
19. Hawkes, R. <u>Anal. Biochem.</u>, 1982, <u>123</u>, 143.
20. Maxam, A.M.; Gilbert, W. <u>Methods Enzymol.</u>, 1980, <u>65</u>, 499.
21. Palmenberg, A.C.; Kirby, E.M.; Janda, M.R., Drake, N.L.;Duke, G.M.; Potratz, K.F.; Collett, M.S. <u>Nucleic Acids Res.</u>, 1983, <u>12</u>, 2969.
22. Vandeyar, M.A.;Weinder, P.W.; Hutton, C.J.; Batt, C.A. <u>Gene</u>, 1988, <u>65</u>, 129.
23. Stueber, D.; Bejard, A. <u>EMBO J.</u>, 1982, <u>1</u>, 1399.
24. Nakamura, K.; Inouye, M. <u>EMBO J.</u>, 1982, <u>1</u>, 771.
25. Haschemeyer, R.H.; Haschemeyer, A.E.V. In <u>Proteins</u>; Wiley-Inter Sci. Publ.: 1973; p 89 - 90.
26. Castle, A.V.; Wheelock, J.V. <u>J.Dairy Res.</u> 1972, <u>39</u>, 15.
27. Jimenez-Flores, R.; Kang, Y.C.; Richardson, T. In <u>Protein Tailoring for Food and Medical Uses</u>; Feeney, R.E. and Whitaker, J.R. Eds.; Marcel Dekker, Inc., 1986; p155.

RECEIVED May 16, 1990

Chapter 15

Rheology

A Tool for Understanding Thermally Induced Protein Gelation

D. D. Hamann

Food Science Department, North Carolina State University,
Raleigh, NC 27695-7624

General concepts of rheology are discussed as applied to the protein gelation process and the texture of the final gel. Small strain oscillatory experiments are presented as a means of studying the gelation process. Large strain (fracture) rheology is presented as a means of studying the fundamental mechanical properties of the final gel. Examples of oscillatory testing show rheological changes during heating and cooling thermal scans and isothermal heating. Fracture testing examples show effects of fillers and enzymatic hydrolysis.

Many foods are formed by the gelation of proteins. This gelation transforms the material physically from a viscous sol or liquid into a solid which is quite elastic in its response to a physical force application. Rheology has been defined as "the science of deformation of matter" (1) and is therefore an appropriate tool for observing physical characteristic changes influenced by variables such as time, temperature and pH or various combinations.

Rheology as used by polymer chemists and others in the study of non-food materials has usually been limited to materials that flow (do not exhibit catastrophic fracture) or small strain studies of viscoelastic solids (strains are smaller than those producing catastrophic fracture). Whorlow (2) discusses the narrowing of the meaning of the term rheology. When fracture is of interest, it is usually studied in a separate area called fracture mechanics. Emphasis is on the study of crack extension as a function of applied forces (3). Linear fracture mechanics is a term used when large plastically yielded regions surrounding cracks or flaws are not present.

The separation of the study of food deformation into rheology and fracture mechanics does not seem practical because a primary concern is how food flows or moves during mastication which, in the case of a solid, is a function of how it fractures and breaks down. There are relatively few food rheologists and most of these must, of necessity, work with the full range of foods, liquid to solid. The food rheologist must base studies on the physics of flow, small strain deformation, and fracture mechanics and it is desirable they be familiar with the psychorheology of sensory food evaluations (4).

0097–6156/91/0454–0212$06.00/0

Typical food protein gels include frankfurters and other comminuted meat products, egg products, shellfish analogs from surimi, dairy protein gels, and plant protein gels. Some of these will be used as examples in this chapter.

The objective of the chapter is to demonstrate how food rheology can be used to better understand gelation transitions involved in product manufacture and how material properties, quantified in fundamental units, relate to sensory quality.

Sol-Gel Transition Rheology

Background. Modern rheometers distinguish between viscous and elastic resistance to forces as well as quantify the magnitudes of resistance, deformation, and rate of deformation. Figure 1 represents a deformed element of material originally a rectangular parallelepiped of length (L) and height (h). The top has been pushed to the right by a tangential force (F) so that the side view is now a parallelogram rather than a rectangle. This type of deformation changes shape but does not change the volume of the material element. It is called shear and is the only type of deformation that will be discussed in this chapter. If the motion of the top surface is continuous we have the situation common to fluids tested in viscometers. The shearing surfaces may be curved in various ways but the shearing action is always present. Shear strain (γ) is defined as

$$\gamma \approx \tan \gamma = \Delta L/h \tag{1}$$

The rate at which this angle changes is the shear strain rate (shear rate) and is commonly the controlled variable in viscometry. Shear stress is the applied force divided by the area of the surface to which it is applied tangentially. In Figure 1 the force would be applied to the top surface area and the bottom surface provide an equal magnitude but opposite direction force with the result being the deformation shown.

In the case of fluids behaving in a viscous manner, the viscosity (η) is

$$\eta = [\text{shear stress}] / [\text{shear rate}] \tag{2}$$

For solids, shear strain is assumed small and a shear modulus (G) is defined as

$$G = [\text{shear stress}] / [\text{shear strain}] \tag{3}$$

G is sometimes referred to as the modulus of rigidity.

Common Test Modes. Consider the rheometer in Figure 2. It will operate in several modes including:
   (1) Viscometry mode - shear strain rate is increased in increments and the output at each is viscosity, shear stress, and strain rate. The material is assumed to be viscous in this test.
   (2) Flow relaxation - after a specified shear history, the instrument stops its motion (strain rate = zero) and the decay in shear stress is monitored.
   (3) Relaxation - with the material at rest, a specified small strain is imposed and maintained constant. The decay in shear stress is monitored.
   (4) Stress growth - a specified shear strain rate is applied and the shear

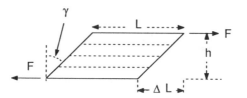

Figure 1. An element of material deformed in shear

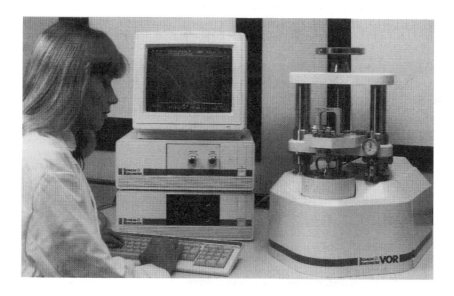

Figure 2. A modern rheometer (courtesy of Bohlin Rheologi)

stress monitored. A typical response is that the stress vs strain rate line is fairly linear to a maximum stress (yield stress) after which the stress drops to a lower stable value.

(5) Strain sweep - sinusoidal oscillatory shear strain is imposed and the resulting shear stress monitored. Frequency is held constant and strain amplitude increased incrementally from a minimum to a maximum or other sequence as desired.

(6) Frequency sweep - this is similar to (5) but the strain amplitude is held constant and frequency varied incrementally.

Superimposed on any of the above tests is a programed temperature history that can imitate a cooking process or other event.

Physical Meaning of Rheological Information. Proper interpretation of results is critical. A brief mathematical description with explanation of physical meaning will be presented for the case of imposed sinusoidal shear strain. Sinusoidal shear strain can be written as

$$\gamma = \gamma_0 \sin(\omega t) \tag{4}$$

where γ_0 is the strain amplitude, ω is the angular frequency in radians/s and t is the time. The shear rate will be the derivative of Equation 4 with respect to time

$$\text{shear rate} = \gamma_0 \omega \cos(\omega t) \tag{5}$$

The shear stress (τ) will, in general, be out of phase from the strain by an angle δ and can be written as

$$\tau = \tau_0 \sin(\omega t + \delta) \tag{6}$$

where τ_0 is the stress amplitude. The ratio of shear stress divided by shear strain can be written as the sum of two components, one in phase with the strain and the other 90° out of phase (5). A complex number coordinate system is often used for mathematical clarity and easy manipulation. The y axis is the imaginary axis and the real part of the number is represented on the x axis. The real term is the in-phase part and the imaginary term is the out of phase part

$$G^* = (\tau_0/\gamma_0) \ [\cos \delta + i \sin \delta] = G' + i \ G'' \tag{7}$$

where i is the imaginary number $(-1)^{1/2}$, G' is the storage modulus, G'' is the loss modulus and G* is the complex modulus. The ratio (τ_0/γ_0) is the absolute modulus, |G*|. The absolute modulus is a measure of the total unit material shear resistance to deformation (elastic + viscous). For a perfectly elastic material the stress and strain are in phase ($\delta = 0$) and the imaginary term is zero. In the case of a perfectly viscous material $\delta = 90° = \pi/2$ radians and the real part is zero. The ratio G''/G' is called the loss tangent and is equal to the tangent of the phase angle

$$\tan \delta = G''/G' \qquad (8)$$

This is proportional to the (energy dissipated) / (energy stored) per cycle. Specifically, the energy dissipated due to viscous behavior per cycle divided by the maximum elastic energy stored is $2\pi \tan \delta$ (6). A perfectly elastic material would exhibit $\delta = 0$ whereas for a perfectly viscous material $\delta = 90°$. In this later case $G' =$ zero and all work of deformation is converted to heat raising the temperature of the material. Protein gels are normally quite elastic so values of δ are near 10°. A muscle sol prior to cooking exhibits some elasticity but is viscous so a typical δ would be 45°. The transition from a sol to a gel is very evident from changes in δ.

Most food gelation studies have been done at a constant ω (often < 1 Hz) with temperature and/or time varying (7-11). Egelandsdal et al., (12) found significant effects on G' and δ of myosin gels when strain amplitude was varied from 0.003 to 0.1. It is necessary to know and record instrument controlled variables including oscillation frequency and strain amplitude.

An Example - Conalbumin. A relatively straightforward rheological study of gelation can be done using an oscillatory test mode holding frequency and strain amplitude constant while varying temperature through a gel forming history. Best results are usually obtained at low frequencies because it is at low strain rates that molecular properties are elucidated (13, 14). Figure 3 shows heating and cooling response curves for conalbumin protein. Below about 62°C the conalbumin solution behaved primarily as a low viscosity liquid and the torque sensing element used was not able to detect fluid resistance to shear above instrument noise. At about 62°C it could be seen that the phase angle was low, about 5°, and tending to decrease. This indicated elastic response was dominant suggesting an elastic structure was forming, but it was not very rigid since G' was very low. At about 63°C, G' started to increase rapidly indicating the elastic structure was being stiffened, probably by additional bonds being formed. This continued to near 75°C at which temperature the rate of increase in G' started to decrease and the phase angle started to increase with the slope increasing as temperature increased. The increase in phase angle indicates more of the strain energy was going into viscous energy loss. This suggests an increase in volume fraction of the protein caused by additional unfolding of the molecules. From 80°C to 95°C, G' was stable but the phase angle continued to increase. Evidently, protein unfolding continued to 95°C. This is different than what has been observed for muscle proteins where the phase angle decreases to a low value where it remains as temperature is increased (e.g., 15).

Upon reaching 95°C, the conalbumin was cooled at 1°C / min to 25°C. Except for a flat region from 95°C to 88°C, G' increased linearly as temperature decreased. This was probably due to increased hydrogen bonding. The phase angle decreased at a steep slope to about 55°C temperature after which it was stable at an angle near 4°. Phase angles for foods do not normally drop much below 4°. The G' flat region, 85°-95°C, is probably controlled by hydrophobic or covalent associations.

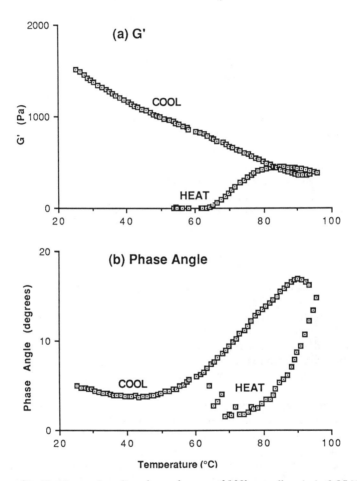

Figure 3.   Heating and cooling thermal scans of 10% conalbumin in 0.85 NaCl solution; scanning rate 1 °C per min

<u>Limitation of the Thermal Scanning Technique.</u> A limitation of the thermal scanning method is that a favorable protein gelation temperature range may not be maintained for an adequate time to fully develop a gel characteristic of that range, particularly if the protein concentration is low (Wu, J.Q. et al. <u>J. Agric. Food Chem.</u>, in review). It has been reported that slow aggregation produces a "more ordered" gel network characterized by higher elasticity (<u>16</u>). Wu et al. (Wu, J.Q. et al. <u>J. Agric. Food Chem.</u>, in review) have shown that isothermal conditions producing myosin gels of high shear modulus and elasticity require a long time for full development (Figure 4) with gel development being delayed by low myosin concentration. This paper also discusses myosin unfolding as the most likely critical factor with lower temperatures being favorable to aggregation and gelation. It is well known that some fish muscle sols gel irreversibly overnight at 0°C (<u>17</u>). Wu et al. (Wu, J.Q. et al. <u>J. Agric. Food Chem.</u>, in review) show that chicken myosin heated very rapidly to temperatures of 74°C or above produced no gel whereas good gels were formed rapidly at lower temperatures, 48°C being near optimum .

With the limitations discussed in mind, thermal scanning rheology is a useful tool for the study of food protein gelation. Changes in the gelation curves can be studied as affected by protein source, protein concentration, specific cooking histories, pH adjustment, additives and many other factors. The phase angle, δ, and storage modulus, G', (or other parameters obtainable from these) are key indicators of gelation occurring as already discussed with respect to Figure 3.

<u>Fundamental Properties of Food Gels Related to Texture</u>

<u>Definitions.</u> Transition rheology helps understand protein gelation chemistry which produces final gel properties. However, it is difficult to consistently predict sensory texture from thermal rheology scans or other physical testing of the gels at strains insufficient to cause gel fracture (rupture). Depending on the literature on the class of materials tested, properties at rupture may be called, ultimate, failure, rupture, fracture or by some other descriptive term. To be consistent, in this chapter we will use the adjective 'fracture' to describe these properties.

<u>Axial Compression.</u> Equations for calculating stresses, strains or moduli from instrumental tests are based on assumptions that may not be valid for the very deformed conditions causing fracture. This is particularly true if an assumption was small deformation. Axial compression of a cylindrical specimen between parallel plates, for example, causes the specimens height to decrease and its diameter to increase. Using the undeformed height and diameter in equations based on the assumption of small deformation is inappropriate if the ratio of decrease in height divided by initial height is larger than about 0.1. Many protein food gels are very deformable and this ratio can approach unity so, if fundamental units independent of a specific test are to be used to specify fracture conditions, equations based on larger strain conditions should be used. A number of approaches to this are in the literature (see <u>18</u>).

One approach to developing an appropriate axial strain equation for compression of cylinders at large axial strain is to assume the basic material properties are independent of the deformation and to consider the appropriate strain to be the sum of an infinite number of small axial strain changes. Integrating these small strains (each one based on the previous specimen height), the result is Hencky's strain and given by

$$\text{Hencky's strain} = - \ln [1- (\Delta h/h)] \qquad (9)$$

If a gel is fairly free of voids (gas space) it will be nearly incompressible (in terms of no volume change). Equation 9, the assumption of incompressibility, and the relationship between axial strain and shear strain (19) can be used to develop the following equation for fracture shear strain:

$$\gamma_f = 1.5\{- \ln [1- (\Delta h/h)]\} \qquad (10)$$

where γ_f is the fracture shear strain. Inserting $\Delta h/h$ values into this equation reveals that a Δh equal to 50% of h produces a γ_f of about 1.0. Thus, a distorted specimen is required to produce a γ_f of 1.0. With the incompressibility assumption the volumes of the undeformed cylinder and the deformed cylinder will be equal and the equation for shear stress can be written

$$\tau_f = 0.5\ F_c/[\pi R^2/(1-\Delta h/h)] \qquad (11)$$

where τ_f is the fracture shear stress, F_c is the uniaxial compression force at fracture, and R is the initial cylinder radius. Many food gels including frankfurters, surimi based seafood analogs, and unwhipped egg white can be considered incompressible so the above equations are suitable. Peleg (20) discusses aspects of the Hencky strain approach applied to food force /deformation curves.

One reason shear stress and strain are calculated is that, experimentally, shear fracture is almost always exhibited by the specimens subjected to axial compression. The maximum shear stresses and strains can be shown to be at an angle of 45° from the cylinder axis which for a perfectly homogeneous specimen would produce conical fracture surfaces. Actual gels tend to fragment somewhat but the 45° angles (based on the deformed cylinder height) are usually very evident. Compression failure is resisted by gel incompressibility. Rarely, fracture may occur as a vertical crack in the surface of the cylinder. Culioli and Sherman, (21) show this is common for cheese. This is a tension fracture and can be thought of as the type of failure prevented in a barrel by hoops.

Other assumptions are made in developing Equations 10 and 11 which one needs to be aware of. It is assumed that the deformed shape is still that of a cylinder, not a barrel shape or a hourglass shape. It is also assumed that the friction between the plates and the specimen as R increases can be neglected. Actually, these are not independent. If friction is high, a barrelling will occur and if friction is very low a hourglass shape will develop. The plates may have to be lubricated so the deformed specimen cross-section is of approximately constant area throughout its length. Experimentally, the length to diameter ratio can have an effect on results, particularly if the ratio is less than about 0.7 (20, 22, 23,). In the references cited in this chapter, the length to diameter ratio ranged from 0.7 to near 2.0.

It would seem from all the considerations discussed above that a uniaxial compression test may not be the best test for obtaining data on fracture stress and strain conditions in terms of fundamental units. Although this is probably the case, it is an easy test to perform and collect data. If care is taken it seems that valid basic information can be obtained up to shear strains somewhat over 1.0. Montejano et al. (24) evaluated fracture of fish surimi gels at room temperature using axial compression and compared results with data from a torsion method finding results from the two methods in general agreement. Shear strains below 1.1 were not significantly different (p>0.05). For shear strains much above 1.0, however, a better method is needed.

An approach similar to that above can be used for axial tension (18). Because of sample attachment difficulties, tension has not been used as frequently as axial compression. This may change as some ingenious attachment methods have been described recently (e.g., 25).

Torsion Testing. Many of the uncertainties in calculating fundamental fracture parameters can be eliminated by going to a torsion (twisting) test. Consider the following advantages:

(1) This test produces what is called pure shear; a stress condition that does not change the specimen volume even if the material is compressible..
(2) The specimen shape is maintained during the test minimizing geometric considerations.
(3) Because of (1) and (2), the calculated shear stresses and strains are true values up to large twist angles (45°, equivalent to shear strains near 1.0).
(4) There is no restriction on the criterion for fracture. The material can fail in shear, tension, compression, or a combination mode.
(5) Principle (maximum) shear, tension, and compression stresses all have the same magnitude but act in different directions so it is easy to determine if the material failed due to shear, tension, or compression.
(6) Friction between the specimen and test fixture does not have to be considered.

Several assumptions are made in developing the equations for torsion of specimens with circular cross sections (see any strength of materials book; e.g., 26) which have been shown to be valid at small strains. Based on the work of several authors (27-30), Diehl et al. (22) presented a method for torsion testing of solid homogeneous foods extending the applicable strain range much higher. It was later shown that protein gels tested in this manner produced results that correlated with sensory texture profiling (31). Fracture shear stress correlated best with sensory hardness and fracture shear strain with sensory cohesiveness. This was confirmed for surimi gels (32). From experimental results it seems that the shear strain limit for the torsion test is when the specimen shape at the critical cross section noticeably changes. For surimi gels this can be at a shear strain as high as 3.0 (32, 33).

A specimen ready for twisting in a Brookfield viscometer, model 5xHBTD (Brookfield Engineering Laboratories, Inc., Stoughton, MA) is shown in Figure 5. In

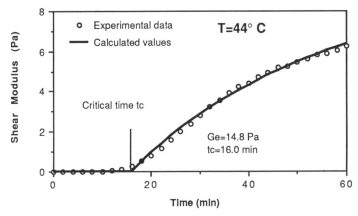

Figure 4. *Shear modulus development of chicken myosin at 44° C; protein concentration 10 mg/mL (Reprinted from Wu, J. Q. et al. J. Agric. Food Chem., in press. Copyright 1991 American Chemical Society).*

Figure 5.   *Placing the torsion specimen in the test instrument*

the example that will be used here specimens are prepared from cylindrical specimens (length = 2.87 cm, diameter approximately 1.9 cm) by gluing a disposable notched styrene disk on each end (cyanoacrylate glue), placing in a machining instrument, and machining an annular groove around the cylinder. The disks also serve to hold the specimen in the test viscometer and transmit the twisting rotation to the specimen. The annular groove insures that fracture occurs at the base of the groove not affected by the glued ends or sharp angles between specimen surfaces. Geometric factors can be included in equation constants producing fairly simple algebraic equations for calculating fracture shear stress and strain.

Gels are known for elastic behavior and this is generally true for protein gels. Torque vs angle of twist graphs are normally quite linear. With the assumption of a linear torque vs angle of twist, a rotational rate of 2.5 rpm, a groove width of 1.27 cm and a 1.0 cm cross-section diameter at the center of the groove, typical equations for fracture shear stress and uncorrected shear strain (γ) are

$$\tau_f, Pa = 1580 \text{ x the torque in instrument units} \tag{12}$$

$$\gamma, dimensionless = 0.150, s^{-1} \text{ x time to fracture, } s$$
$$-0.008848 \text{ x the torque in instrument units} \tag{13}$$

The last term in Equation 13 is due to spring windup in the 5xHBTD viscometer. For an instrument with a very stiff sensing element this term would not be present. If shear strain is above about 1.0, γ should be corrected using an equation given by Nadai (29). Returning to the notation that γ_f is the true shear strain of interest

$$\gamma_f = \ln[1 + \gamma^2/2 + \gamma(1 + \gamma^2/4)^{1/2}] \tag{14}$$

A fractured surimi gel specimen is shown in Figure 6. In contrast to axial compression testing, a diagonal crack in torsion indicates failure in tension (34). If the break was perpendicular to the cylindrical axis of the specimen it would be a shear fracture.

Saliba et al. (35) used axial compression and torsion to evaluate frankfurter fracture. At torsion shear strains of 1.45 axial compression results were about 13% higher for both stress and strain (note that $\Delta h/h$ was 0.62 so the original cylinder was grossly deformed). A possible factor in this study was that the torsion test involves the interior of the original specimen whereas the axial compression of unmachined cylindrical specimens involved material closer to the cooking tube surface. If radial nonhomogeneity is present, material at the largest test cross-section radii will have the greatest influence. Unpublished results for frankfurters from the authors laboratory suggest that, if similar cross section locations are used in both axial compression and torsion testing, shear stresses agree quite closely if torsional γ_f is below about 1.4 but axial compression γ_f values are somewhat lower than 1.4. For larger strain conditions, typical of chicken frankfurters and surimi based gels, the axial compression τ_f is larger than the torsion value and the axial compression γ_f is smaller than the torsion value. From comparisons made to date it seems that protein gel fracture shear stress and strain data from axial compression tests (usually with the ends lubricated) and torsion agree quite well up to shear strains of about 1.0 and do not become greatly different up to shear strains of about 1.4. More work needs to be done on this.

Example application of shear stress and shear strain data. An example of how fracture stress and strain values can be used effectively will be based on Figure 7. Each data point in Figure 7 is the mean of 2 replications tested in torsion at room temperature using the methodology compatible with Equations 12-14. Twenty three different commercial labels were tested with 10 specimens tested per replication (unpublished data). The geometric shapes are drawn to show the grouping of the data. Notice that there is a general linear relation between stress and strain with all data considered. It can then be seen that pork/beef combinations (all were 30% fat) were at the low stress/strain end and chicken frankfurters (about 20% fat) were at the high end. Beef franks were at about the same strain level as pork/beef combinations but tended to be higher in stress. Turkey (20% fat) franks had about the same stress level as beef franks but were higher in strain. In general, the commercial poultry franks exhibited higher strain values than the commercial red meat products. All of the observations above were highly significant statistically. Sensory texture profile development for each of the products was consistent with the instrumental results. Specifically, the chicken and turkey products were the most cohesive (deformable) and the red meat products the least cohesive.

Figure 7 is a tentative base for adjusting textural attributes. If one desires to make a chicken frank with the texture of a pork/beef frank it is obvious that the fracture shear strain must be decreased and also the shear stress. Protein concentration, thermal process, and a number of other modifications can be used to adjust shear stress but shear strain is harder to change (36, 37). The base gel seems to be the primary factor affecting fracture strain and this gel must be interfered with or enhanced as the case may be (33). Recent unpublished work in the authors laboratory used Figure 7 as base information for evaluating the suitability of a proposed fat substitute in frankfurters. Both red meat and poultry applications were tested and it was possible to keep red meat at a strain of about 1.3 and move chicken franks to near the red meat data and at the same time reduce both the fat content and calorie level. Sensory profile evaluation again confirmed the instrumental results.

A Second Example. This example is taken from Hamann et al. (33) and demonstrates the effects of a change in the base gel and/or a filler ingredient. Figure 8 shows how two variables, protease inhibition and/or starch addition, influenced the fracture of gels made from a specific lot of menhaden (Brevoortia tyrannus) surimi. Surimi made from this species (and many others) contains a heat stable alkaline protease which causes gel degradation at temperatures near 60°C (38). Hamann et al. (33) give evidence that beef plasma or egg white will inhibit the enzymatic breakdown of the protein. Figure 8 shows fracture results for 9 % protein, 2% NaCl surimi gels formed by thermally processing at 60°C for 30 min followed by 15 min at 90°C. The inhibitors increased the

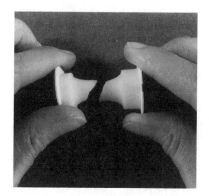

Figure 6. Surimi gel specimen showing the fracture

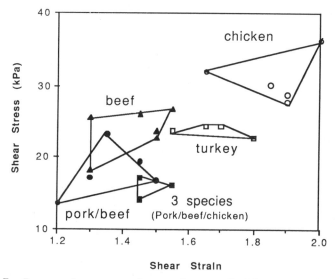

*Figure 7. Fracture shear stress - shear strain means for 23 commercial frankfurters
(23 different labels)*

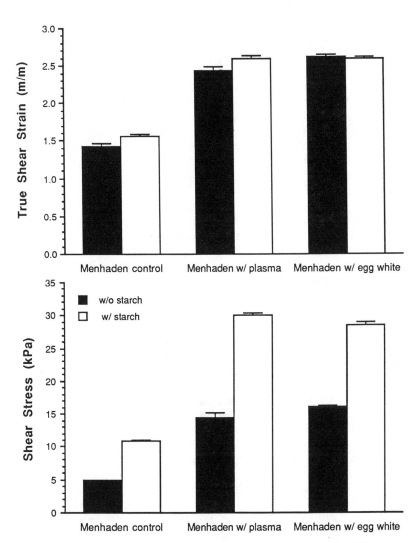

Figure 8. The effects of protease inhibitors (plasma or egg white) and 5% unmodified potato starch on fracture shear stress and strain of menhaden surimi gels cooked at 60 °C for 30 min followed by 90 °C for 15 min (Reproduced with permission from ref. 34. Copyright 1990 Institute of Food Technologists.)

fracture shear strain by about 60-70% and tripled the fracture shear stress. The second
variable, adding 5% potato starch, doubled the shear stress but did not affect the shear
strain significantly. These results are consistent with results from earlier work (37, 39)
showing that proteolytic activity affects the base gel while starch acts as a filler
influencing primarily gel hardness. Hamann and Lanier (32) show that protein dilution
by adding water to a surimi gel lowers stress but has a very limited effect on fracture
strain.

Some general observations are that:

(1) Changes in the base gel result in changes in both fracture stress and
strain (in the above example the protease degraded the base gel).

(2) Non reactive fillers, protein dilution, or other changes that do not change
the base gel affect primarily fracture stress (the starch increased the fracture
stress but not the fracture strain; added water decreases stress).

Concluding Remarks

This brief chapter with examples gives some insight into the use of rheology to better
understand gel formation and gel quality. Articles on the subject are becoming much
more numerous and rheology combined with companion tools such as scanning
electron microscopy and differential scanning calorimetry will enable us to probe the
intricacies of protein gels.

Literature Cited

1. Muller, H.G. An Introduction to Food Rheology; Crane, Russak & Co.: New
York, 1973; p 1.
2. Whorlow, R.W. Rheological Techniques; Ellis Horwood: Chichester, UK, 1980; p
17.
3. Kobyayashi, A.S. Experimental Techniques in Fracture Mechanics; Monograph
No. 1, Soc. for Exp. Stress Anal.: Westport, CN, 1973; p 4.
4. Frijters, J.E.R. In Sensory Analysis of Foods; Piggott, J.R., Ed.; 2nd ed.;
Elsevier: Barking, Essex, England, 1988; p 131.
5. Dealy, J.M. Rheometers for Molten Plastics; Van Nostrand Reinhold: New York,
1982; p 51.
6. Whorlow, R.W. Rheological Techniques; Ellis Horwood: Chichester, UK, 1980; p
250.
7. Beveridge, T.; Jones, L.; Tung, M.A. J. Agric. Food Chem. 1984, 32, 307.
8. Bohlin, L.; Hebb, P.; Ljusberg-Wahren, H. J. Dairy Sci. 1984, 67, 729.
9. Beveridge,T.; Timbers, G.E. J. Texture Stud. 1985, 16, 333.
10. Samejima, K.; Egelandsdal, B.; Fretheim, K. J. Food Sci. 1985, 50, 1540.
11. Noguchi, S.F. Bull. Jap. Soc. Sci. Fish. 1986, 52, 1261.
12. Egelandsdal, B.; Fretheim, K.; Harbitz, O. J. Sci. Food Agric. 1986, 37, 944.
13. Bohlin, L; Egelandsdal, B.; Martens, M. In Gums and Stabilizers for the Food
Industry 3; Phillips, G.O.; Wedlock, D.J.; Williams, P.A., Eds.; Elsevier:
Barking, Essex, England, 1986; p 111.
14. Hamann, D.D.; Purkayastha, S.; Lanier, T.C. In Thermal Analysis of Foods;
Harwalker, V.R.; Ma, C.Y., Eds.; Elsevier: Barking, Essex, England, 1990;
p 330.

15. Sano, T.; Noguchi, S.F.; Matsumoto, J.J.; Tsuchiya, T. J. Food Sci. 1990, 55, 51.
16. Gossett, P.W.; Rizvi, S.S.H.; Baker, R.C. Food Technol. 1984, 38 (5) 67.
17. Ikeuchi, T.; Simidu, W. Bull. Jap. Soc. Sci. Fish. 1963, 29, 151.
18. Peleg, M. J. Texture Stud. 1985, 16, 119.
19. Polakowski, N.H.; Ripling, E.J. Srength and Structure of Engineering Materials; Prentice-Hall: Englewood Cliffs, NJ, 1966; pp. 57-60.
20. Peleg, M. J. Texture Stud. 1977, 8, 282.
21. Culioli, J.; Sherman, P. J. Texture Stud. 1976, 7, 353.
22. Diehl, K.C.; Hamann, D.D.; Whitfield, J.K. J. Texture Stud. 1979, 10, 371.
23. Chu, C.F.; Peleg, M. J. Texture Stud. 1985, 16, 451.
24. Montejano, J.G.; Hamann, D.D.; Lanier, T.C. J. Rheology. 1983, 27, 557.
25. Langley, K.R.; Millard, D.; Evans, E.W. J. Dairy Res. 1986, 53, 285.
26. Polakowski, N.H.; Ripling, E.J. Srength and Structure of Engineering Materials; Prentice-Hall: Englewood Cliffs, NJ, 1966; p 377.
27. Davis, E.A. J. Appl. Physics. 1937, 8, 231.
28. Morrison,J.L.M. Proc. Inst. Mech. Engrs. 1948, 159, 81.
29. Nadai, A. J. Appl. Physics. 1937, 8, 205.
30. Neuber, H. AEC Translation Series, AEC-tr-44547 , 1958
31. Montejano, J.G.; Hamann, D.D.; Lanier, T.C. J. Texture Stud. 1985, 16, 403.
32. Hamann, D.D.; Lanier, T.C. In Seafood Quality Determination; Kramer, D.E.; Liston, J., Eds.; Elsevier: Barking, Essex, England, 1987; p 123.
33. Hamann, D.D.; Amato, P.M.; Foegeding, E.A. J. Food Sci. 1990, 55, 665.
34. Hamann, D.D. In Physical Properties of Foods; Peleg, M.; Bagley, E.B., Eds.; AVI: Westport, CN, 1983; Chapter 13.
35. Saliba, D.A.; Foegeding, E.A.; Hamann, D.D. J. Texture Stud. 1987, 18, 241.
36. Wu, M.C.; Hamann, D.D. J. Texture Stud. 1985, 16, 53.
37. Hamann, D.D. Food Technol. 1988, 42 (6), 66.
38. Boye, S.W.; Lanier, T.C. J. Food Sci. 1988, 53, 1340.
39. Kim, B.Y.; Hamann, D.D.; Lanier, T.C.; Wu, M.C. J. Food Sci. 1986, 51, 951.

RECEIVED August 14, 1990

Chapter 16

Food Dough Constant Stress Rheometry

Jimbay Loh

Kraft General Foods, 555 South Broadway, Tarrytown, NY 10591

Constant stress rheometry (creep) is a rheological
technique which can provide useful information on
stress-controlled effects of bread dough. Unlike
the "spread test", creep can quantify fundamental
rheological parameters that characterize the
stress-strain-time behavior of the bread dough.
Using quantified parameters (i.e. instantaneous
compliance, retarded compliance, Newtonian
viscosity and retardation time), the effects due
to various ingredient addition and the degree of
mixing were evaluated.

For baked goods, the consistency and the rheology of dough have
profound effects on its handling properties and on the textural
qualities of the finished products (1). Traditionally, the
rheology of food dough was characterized using either empirical or
fundamental measurements. The former measurement employes a complex
mode of force loading (e.g. Mixograph) and/or an ill-defined sample
geometry (e.g. spread test). Consequently, the obtained results are
impossible for interpretation in basic physical terms. On the
contrary, the later measurement uses simple mode of force loading
(e.g. shear, compression or tension) and a well-defined sample
geometry. It also takes the time factor or shear rate into
consideration, thus is more suitable for the characterization of
viscoelastic food dough.
    Three fundamental measurements are commonly used, namely
dynamic oscillation, stress relaxation and creep. Most available
rheometers for characterizing the viscoelastic materials using
dynamic oscillation method are applicable over rather restricted
ranges of the time and frequency domains (typically from 0.01 to 10
Hz). In the case of thermorheological complex materials (e.g. food
dough) such limitation can not be circumvented by temperature
superposition. With stress relaxation and creep testing, a time/

0097–6156/91/0454–0228$06.00/0

shear rate range of 6 to 8 decades is possible. Among the three
fundamental measurements, creep measurement is perhaps the least
popular test used in the United States for characterizing bread
dough. Nevertheless, creep tests have been widely used in studying
food systems in Europe. It is particularly suitable for studying
the rheological phenomenon of food dough that is basically stress
driven rather than shear rate controlled (e.g. spreading and rising
of bread dough during proofing and baking). It is also a powerful
tool capable of illustrating rheological structure. The creep
function (strain vs. time relationship) approximates a mirror image
of the relaxation function (stress vs. time relationship) reflected
in the time axis. The creep compliance (strain divided by stress)
also approximates the reciprocal of the relaxation modulus (stress
divided by strain). However, creep function reveals a somewhat
simpler picture than relaxation function and perhaps can be more
easily related to the functionality of dough. With computer-aidded
data analysis of modern control stress rheometers, creep parameters
can be readily calculated. The flow behavior of food dough can be
expressed in terms of a rheological model in which the elastic
elements are represented as springs and viscous elements are
represented as dashpots. Reiner (2) used a five-element model for
bread dough (Figure 1) to show yield value with a frictional element
(A), instantaneos elasticity with an elastic element (B), Newtonian
viscosity with a dashpot (C) and retarded elasticity and retarded
viscosity, respectively, with an elastic element (D) and a dashpot
(E) in parallel. Using a six-element model (Figure 2), Shama and
Sherman (3) studied ice cream and suggested that instantaneous
elasticity (E_o) is affected primarily by ice crystals, retarded
elasticity (E_1) and retarded viscosity (n_2) by the weak
stabilized gel network, retarded elasticity (E_2) by protein-
enveloped air cells, retarded viscosity (n_2) by fat crystals, and
Newtonian viscosity (n_N) by both fat and ice crystals. Using a
four-element Voigt model (Figure 3), the present study takes an
similar approach of Shama and Sherman for ice cream, to determine
the qualitative contribution of selected bread ingredients to the
specific model constants for the bread dough. This 4-element model
can be expressed as follow:

$$J(t)=J_o+J_R[1-exp(-t/T_R)]+t/n_N$$

where $J(t)$ is the total strain devided by stress at time t; J_o is
instantaneous compliance, J_R is retarded compliance, T_R is
retardation time which equals to J_R x n_R; n_R is retarded
viscosity; and n_N is Newtinian viscosity.

The existing theories have been developed for homogeneous,
linear viscoelastic materials. The difficulty in developing and
using the constitutive equation for complex, nonlinear viscoelastic
food dough is well known. In addition, the dynamic nature of food
doughs (e.g. oxidation, enzymatic reactions, gas formation, moisture
redistribution, microbial fermentation, etc.) renders such an
approach unpractical. The primary objective of this study is to
examine the relationship (if any) between creep parameters of bread
dough and the crumb texture of baked bread. The generated
information was expected to aid in the future data interpretation.

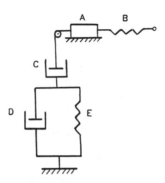

Figure 1. Five element rheological model for the behavior of dough.

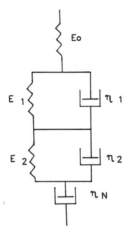

Figure 2. Six element rheological model for the behavior of ice cream.

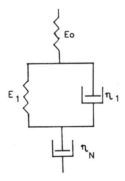

Figure 3. Four element rheological model for the behavior of dough.

In case of the creep test applied to bread dough, the non-
ideal behavior of the dough was first described as a stress-
softening phenomena by Schofield and Scott Blair (4). This implies
a decrease in viscosity with increasing stress. Without an
acceptable theory, applied rheologists often rely on apparent
correlations of measured rheological parameters to the dough
functionality or finished product texture as a partial basis for
data interpretation. Regardless of the experiential and somewhat
unscientific nature of the approach, such correlations greatly
increase the utility of fundamental rheological data in most
industrial environments. Effort was made in the present study to
relate the model constants derived from the creep function of bread
dough to the extensibility, strength and toughness of the finished
bread crumb.

Materials and Methods

Dough Preparation and Testing. All dough samples were prepared
from a commercial patent flour containing 11.8% protein and 12%
moisture. Full formula doughs (Control A) consisted of the
following ingredients (flour basis): 100% of flour, 6.67% of
granular sucrose, 1.67% of salt, 3.33% of nonfat dry milk, 5% of
shortening, 1% of active dry yeast, 0.33% of yeast food, 0.33% of
malt and 61.67% of water. Dough samples were prepared using
National Brand Bread Bakery (Model SD-BT2N) by Matsushita Electric
Trading Co. Ltd. Central Osaka, Japan. These units automatically
form and knead the dough, add dry yeast, proof the dough and bake
the bread. Their precise action cycle is given in Figure 4. The
whole process takes exactly 4 hours to complete. Total six
individual units were used in this study. The variations within
and between different units were found to be less than 7% of the
mean based on the measured loaf volume and crumb texture of finished
bread. Dough samples were obtained by briefly interrupting the
action cycle (see Figure 4). and cutting a small piece of dough
sample (approximately 5 g) using a pair of surgical scissors. Dough
sample was immediately transfered to a constant stress rheometer
(Carri-Med model CS 50; Twimsburg, Ohio) equipped with a 2-cm
parallel plates fixture. Yeasted dough samples were sandwiched
between saran film and punched to a thickness of 1/16 inch to expel
CO_2 before testing. A gap of 500-microns between the plates was
used. After sample loading followed by a 5 minutes of equilibration
time, a constant torque of 2000 dyne.cm was applied to the sample to
obtain the creep data at 25°C. Torque level used in this study
was set just above the yield stress of the stiffest dough sample.
Yield stress of individual dough samples was not measured.
Humidified air and a solvent trap (filled with water) was used to
prevent the sample edge from dehydration during testing. Best fit
creep curve and model constants were calculated based on 4-element
Voigt model. In most cases, use of a six-element Voigt model did
not improve the degree of fit.
     Due to the unstable nature of yeasted dough and the fact that
a considerable portion of the total stress likely underwent

relaxation in a relatively short time, only short time (60 seconds) creep behavior was measured. Under the stress condition used in the present study, an approximate Newtonian flow is normally established after 20 to 35 seconds of stress application thus n_N and other creep parameters can be measured. Both unyeasted (sample taken just prior to yeast addition) and yeast fermented dough samples were tested under the same instrumental conditions.

Full formula dough was used as Control A (see early description for composition). Corresponding variants represent Control A with one single ingredient substracted from the formula (i.e. sucrose, nonfat dry milk, shortening and salt). Similar series was also tested using a reduced formula as Control B by excluding sucrose, nonfat dry milk, shortening and salt from the original formula of Control A. It is apparent that Control A represents a relatively "richer" formula than Control B. Then through the addition of a single ingredient to the Control B, the effects of individual ingredient were determined. In addition, under (less 20 minutes from the control mixing time) and over (plus 30 minutes) mixed dough samples were evaluated.

Confirmatory tests using dynamic measurement were also run for selected dough samples using the identical test geometry and 1% strain. Values of storage modulus, loss modulus, complex viscosity and loss tangent were measured at 1 Hz frequency. Dynamic data were used to check and verify the results derived from the creep data.

Bread Preparation and Testing. All the bread samples of various treatments baked with the Bread Bakery were characterized for specific loaf density and crumb texture. Bread was stored in a brown paper bag overnight. Loaf volume was determined using seeds displacement technique. Loaf density is simply the ratio of loaf weight and loaf volume.

Texture measurement was performed at 20 to 22 hours after baking. A computer-assisted Instron (Model 1122; Canton, Massachusetts) punch and die test was used, adopting the test principle of Dahle and Montgomery (5). Test geometry (Figure 5) was optimized to maximize tensile stress and to minimize compressive and shear stresses by using a 3/4" circular hard aluminum die, a 3/8" rounded brass punch and a crosshead speed of 50 mm/ min. Two slices of bread taken from specific part of each loaf were obtained for testing (Figure 6). A Hobart meat slicer was used to cut the slices from the loaf.

Each single slice was punctured 8 times in the Instron covering the entire slice but avoiding the areas at the very center and near the crust. An average value was calculated and reported. A typical force/ deformation curve and the definitions of the textural parameters measured are given in Figure 7. Two independent parameters, namely crumb strength (F) and crumb extensibility (D) are taken from the curve. A third parameter (i.e. crumb toughness) is calculated by taking the product of F and D. These parameters give better textural description and sample differentiation than compressive firmness measurements using the Baker compressimeter (6) Non-destructive compressive firmness is good for quantifying the bread firming with age, but often fails in differentiating the fresh bread crumbs of drastic textural difference.

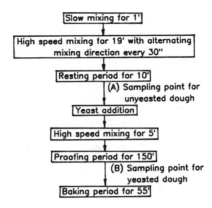

Figure 4. Mixing and baking cycle of National Bread Bakery, showing the sampling point for unyeasted dough (A) and yeast fermented dough (B).

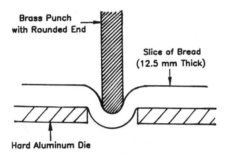

Figure 5. Test arrangement of Instron punch and die test.

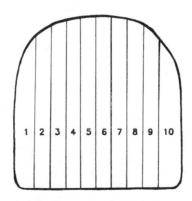

Figure 6. Sampling procedure for bread. Slices no. 3 and 4 were used for the textural characterization of bread crumb.

Crumb strength and extensibility reflect, respectively, sensory firmness and cohesiveness as the crumb is masticated. A separate study (unpublished) indicates that instrumental toughness (FxD) at the range of low toughness, is nearly linear related to the log of sensory toughness. At the very high range of toughness, the differentiation ability of the sensory panel diminishes whereas the instrumental method remains applicable (Figure 8). Using the Instron method, sensory toughness of the bread crumb can be approximated by the instrumental toughness using the following equation:

Sensory Toughness = K (Instrumental toughness)^2

where K is a constant determined by the particular sensory rating scale used.

Data Analysis. All results reported represent the average value of duplicated observations. Typical coefficient of variation (C.V.) between duplicates varies. For example, the C.V. of textural measurement of bread crumb usually ranges from 0.02 to 0.10. But the C.V. of creep and oscillation measurement of the dough often varies more greatly from 0.10 to 0.35, probably due to the dynamic nature of the dough sample. All treatments were randomized. Comparisons made are directional and qualitative. Statistical and quantitative interpretation of the results were not intended.

Results and Discussion

Ingredient Effect on The Creep Parameters of The Dough. Table I and II show the quantified creep parameters of the various unyeasted and fermented dough samples, respectively, through selected ingredient substraction(-) and addition(+).

The data indicate that sucrose weakens the gluten structure of the bread dough as evident by large increase in compliance and decrease in viscosity. Since sucrose has greater hydrophilic properties than most of flour components, it tends to inhibit gluten hydration and dough development. Based on the data in Table I and II, the effect of sucrose can not be fully explained by the increase in fermentable sugar or increased yeast activity. The same softening effect of sucrose has been observed in both yeasted and unyeasted dough.

Addition of nonfat dry milk (NFDM) or shortening results in a decrease in Newtonian and retarded viscosity of fermented dough. NFDM addition weakens the instantaneous elastic element in unyeasted dough. Wheat flour dough can be viewed as a moisture limitted system in which various hydrophilic components compete for hydration. NFDM is mainly made of soluble matters (e.g. proteins and lactose) that absorb moisture rapidly. The observed effect of NFDM on dough rheology can be at least partially explained by the lower degree of gluten hydration. Possible chemical interactions or physical interference between milk proteins and gluten are also possible. It is interesting that the effect due to NFDM on instantaneous

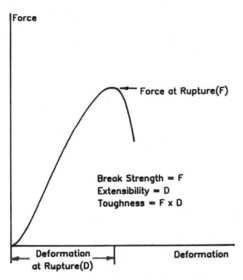

Figure 7. Typical Instron force and deformation curve of bread crumb, showing definition of quantified parameters.

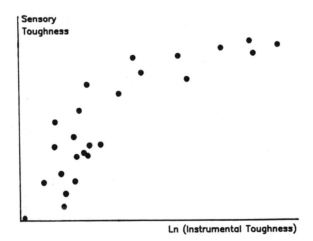

Figure 8. Relationship between instrumental and sensory toughness of bread crumb.

elasticity are different depending on overall bread formula (i.e. rich vs. lean).

Results also suggest a weakening effect due to shortening addition based on the observed increase in instantaneous compliance. Similar to the effect of NFDM, the influence of shortening on retarded compliance is not consistent and depends on whether the shortening was added or substracted suggesting possible interfacial interactions involving NFDM component(s), flour components and shortening.

Table I. Ingredient Effects on The Creep Parameters
of Unyeasted Bread Dough

Samples	J_o	n_N	T_R	J_R	n_R
	(10^{-6}cm²/dyne)	(10^6cps)	(sec)	(10^{-6}cm²/dyne)	(10^6cps)
Control A	60.0	11.9	4.9	15.1	10.6
-Sucrose	5.6	26.8	3.3	29.9	5.5
-NFDM	3.0	70.9	4.2	7.1	5.9
-Shortenin	48.1	13.4	4.5	39.9	1.1
-Salt	84.4	9.3	4.4	22.9	7.3
Control B	45.0	29.1	3.8	20.8	1.8
+Sucrose	591.0	7.1	7.2	41.3	1.7
+NFDM	294.1	8.5	2.4	29.0	0.8
+Shortening	435.0	12.1	4.5	36.8	1.2
+Salt	42.3	33.4	4.3	34.9	1.2

Table II. Ingredient Effects on The Creep Parameters
of Yeasted Bread Dough

Samples	J_o	n_N	T_R	J_1	n_1
	(10^{-6}cm²/dyne)	(10^6cps)	(sec)	(10^{-6}cm²/dyne)	(10^6cps)
Control A	15.6	20.7	4.2	102.0	3.0
-Sucrose	0.2	368.2	4.3	22.2	21.3
-NFDM	3.1	114.1	3.9	36.0	11.0
-Shortening	5.7	51.0	4.5	158.1	12.6
-Salt	27.5	62.5	5.1	141.0	7.3
Control B	38.1	24.5	5.2	105.2	7.8
+Sucrose	340.0	17.3	8.0	247.0	3.2
+NFDM	37.5	5.3	3.7	501.1	0.7
+Shortening	68.8	7.2	4.2	493.3	0.8
+Salt	26.6	31.1	4.1	296.0	1.4

Substraction of salt from the yeasted dough makes the dough less elastic and more viscous. The effect of salt addition in a leaner formula (i.e. ingredient substraction series) gives a dough that is more elastic and also more viscous. It is generally believed that electrostatic attractions between gluten molecules under normal pH are of little significance. However, a stiffening

and a drying (reduced dough stickiness) effect due to salt addition
are also commonly known by the baker. The exact mechanism of salt
affecting dough rheology has not yet been satisfactorily explained.
In general, the effect of salt on creep parameters is relatively
insignificant comparing to NFDM, shortening, etc. except that salt
addition appears to weaken the retarded elasiticity in unyeasted
dough and retarded viscosity in fermented dough.

Minor discrepencies in the data pertinent to the rheological
effect of specific ingredient may be explained by the compositional
differences between Control A and B. Yeast addition/ fermentation
further complicates the picture. Nevertheless, a general weakening
effect of sucrose, shortening and NFDM and a strengthening effect of
salt can readily be demonstrated by the creep data obtained. These
observations are consistent with the common knowledge of the bakers
and the literature. Figure 9 and 10 schematically summarize the
ingredient effects on the individual rheological elements in both
unyeasted and yeasted dough, respectively.

The data obtained from dynamic measurement also show good
directional agreements with the creep data. However, the same
agreements were not always found on a quantitative basis. This is
presumably due to the high dependence of strain, strain rate and
stress history of the bread dough system.

Ingredient Effect on The Texture of Bread Crumb. Under tensile
stress, structural collapse of bread crumb characterized by the
rupture strength and extensibility, mainly relates to the gluten
matrix. The starch gel network and cellular structure may also play
a role in determining the overall texture. Table III shows the
ingredient effect on the textural parameters of the bread crumb.
Our past experience agrees with the current data that break strength
and extensibility are two independent failure parameters.

Table III. Effect of Ingredient Addition and Substraction
on The Texture of Bread Crumb and Loaf Density

Samples	Strength(F) (g force)	Extensibility(D) (mm)	Toughness(FXD) (g X mm)	Loaf Density (g / cm³)
Control A	227	10.9	2474	0.256
−Sucrose	290	12.0	3520	0.340
−NFDM	146	10.0	1510	0.229
−Shortening	219	14.0	3066	0.292
−Salt	211	9.4	1983	0.268
Control B	240	11.0	2640	0.285
+Sucrose	173	10.0	1730	0.256
+NFDM	273	10.0	2730	0.348
+Shortening	207	9.1	1884	0.291
+Salt	266	10.0	2660	0.298

Crumb strength is increased by NFDM and salt and decreased by sugar
and shortening. NFDM is used in bread for nutritional enrichment
and/or control of the color of bread crust. Its effect on crumb

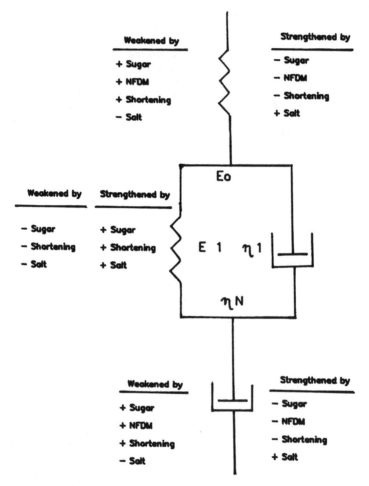

Figure 9. Ingredient effect on the structural elements/ creep parameters of unyeasted bread dough based on four element Voigt model.

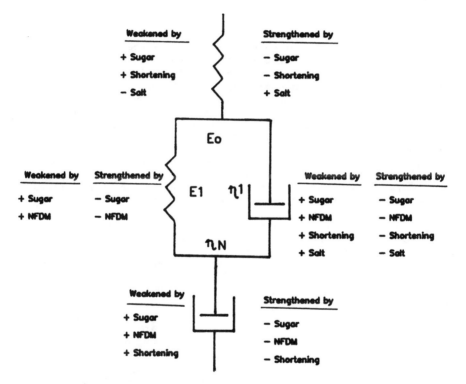

Figure 10. Ingredient effect on the structural elements/ creep parameters of yeast fermented dough based on four element Voigt model.

strength is probably due to reduced free water. Whether the thermal
denaturation of whey proteins in NFDM having a structural effect on
bread crumb or not remains to be determined. Extensibility is
reduced by removing salt from a rich formula (i.e. substraction
series), or by adding sugar, NFDM, shortening or salt to a lean
formula (i.e. addition series). The mechanism of salt induced
change in crumb extensibility is not clear. Crumb toughness can be
reduced by substracting NFDM or salt, or by adding sugar or
shortening to the bread formula. In general, sugar and shortening
decrease the loaf density and NFDM increases the loaf density.

Creep Parameters of The Dough vs. Textural Quality of The Crumb. In
both unyeasted and yeasted systems, no apparent and statistically
significant (p<0.05) linear correlation exist between creep and
textural parameters. Significant correlations (p<0.05) between loaf
density and break strength, and between loaf density and the crumb
toughness were established. No simple linear relation can be
established between loaf density and individual creep parameters.
This is not a total surprise since the structure formation of bread
crumb from the dough is a rather complex process. The rheology of
the dough can only be responsible for part of the variability
observed in crumb texture. Reactions occuring during and after
baking can not be predicted from the dough rheology alone. Based on
the coefficient of determination calculated from stepwisw, multiple
linear regression, Table IV shows the percent variability in crumb
texture or loaf density that may be explained by the measured creep
parameters of the dough. Generally speaking, crumb texture may be
better estimated from the creep parameters of unyeasted dough.
Crumb extensibility and loaf density are the least and most
predictable bread characteristics, respectively.

A positive correlation (r=0.88, p<0.05) between crumb strength
and loaf density suggests that crumb strength is mainly a function
of gas incorporation. No relationship exists between crumb
extensibility and loaf density.

Table IV. Percent Variability in Crumb Characteristics
Explained By The Creep Parameters of The Dough

Crumb Characteristics	Unyeasted Dough	Yeasted Dough
Crumb Strength	76	59
Crumb Extensibility	51	62
Crumb Toughness	78	47
Loaf Density	92	74

Degree of Mixing. The mixing cycle in the "Bread Bakery" was
designed for a specific bread formula recommended by the
manufacturer of the apparatus. The appropriateness of the mixing
condition for the current dough system was assessed by varying
the mixing time and characterizing the resulting dough and bread.
Doughs with 20' reduced and 30' increased mixing time were
evaluated. The results indicate that under or over mixing are

generally undesirable and resulting in a weaker dough (Table V),
weaker crumb and denser loaf (Table VI). This difference is
especially evident in retarded elasticity. Judging from the dough
rheology, crumb texture and loaf density data, it is concluded that
the standard mixing cycle of the "Bread Bakery" unit is suitable for
studying the present experimental system (i.e. Control A). A
maximum dough and crumb strength and a minimum loaf density were
obtained using the standard mixing cycle.

Table V. Effect of Under and Overmixing on The
Creep Parameters of Unyeasted Dough

Samples	J_o	n_N	T_R	J_1	n_1
	$(10^{-6}cm^2/dyne)$	(10^6cps)	(sec)	$(10^{-6}cm^2/dyne)$	(10^6cps)
Undermixed	61.9	9.2	4.8	468.0	1.0
Control A	60.0	11.9	4.9	150.1	1.1
Overmixed	68.1	8.1	3.9	494.2	0.8

Table VI. Effect of Under and Overmixing of The Dough
on The Crumb Texture and Loaf Density

Samples	Strength(F) (g Force)	Extensibility(D) (mm)	Toughness(FXD) (g X mm)	Loaf Density (g / cm^3)
Undermixed	219	9.7	2124	0.320
Control A	227	10.9	2474	0.256
Overmixed	203	9.7	1969	0.269

Conclusions

The major advantage of creep testing of the dough is its capability
of measuring the initial movement of the dough at very low shear
rates. Therefore, the quantified creep parameters can readily be
used to describe the structure of the dough as influenced by the
ingredients and mixing. In practice, the prediction of the crumb
texture and loaf density from the creep parameters of the dough
appears to be limitted even sophisticated statistical models are
used. An attempt was made in the present study to relate the
specific creep parameters to the dough composition.
    Instantaneous elasticity and Newtonian viscosity are reduced
by sugar and shortening addition and enhanced by salt addition.
Increased Newtonian viscosity was observed while NFDM was added.
Ingredient effects on retarded elasticity and viscosity are
inconsistent between unyeasted and yeasted dough, suggesting that
yeast fermentation may cause certain changes in the time dependent
rheological structure of the dough.
    In conclusion, creep testing can be used to characterize the
rheological behavior and structure of food dough. To predict the
final crumb texture, both the dough rheology and the baking
behavior of the dough need to be considered.

Literature Cited

1. Szczesniak, A.S.; Cereal Foods World 1988, 33, 841-843.
2. Reiner, M.; Twelve Lectures on Theoretical Rheology, North-
   Holland, Amsterdam, 1949.
3. Shama, F. and Sherman, P.; J. Food Sci. 1966, 31, 699-706
4. Schofield, R. K. and Scott Blair, G. W.; 1933, Proc. Roy. Soc.
   (London) A141, 72.
5. Dahle, L.K. and Montgomery, E. P.; Cereal Chem. 1978, 55, 197-
   203.
6. American Association of Cereal Chemist. Approved Methods of AACC
   AACC, St. Paul, Minnesota. 1966.

RECEIVED June 20, 1990

Chapter 17

Factors Influencing Heat-Induced Gelation of Muscle Proteins

Denise M. Smith

Department of Food Science and Human Nutrition, Michigan State University, East Lansing, MI 48824-1224

The gelation of muscle proteins is responsible for the textural attributes of many processed meat products. Successful modification of product texture requires an under-standing of the physicochemical or intrinsic properties of proteins and how proteins are influenced by environmental and processing conditions. Percentage of alt-soluble protein and source of the protein (skeletal, cardiac or smooth muscle) in a formulation determine the rheological and microstructural charac-teristics of heat-induced gels. By binding water or interacting with salt-soluble meat proteins the addition of eater-soluble and insoluble proteins may modify gel characteristics. Rheological properties and microstructure of gels are determined by the chem-ical properties of proteins in solution and thus can be modified by changing the pH and salt concentration. The temperatures at which salt-soluble muscle proteins unfold and reaggregate into a cross-linked network may alter gel properties due to interactions with other macromolecules. Non-destructive dynamic rheological properties of salt-soluble proteins can be monitored during a controlled heating process to help elucidate the relationship between the physical and molecular changes during gelation.

The heat-induced gel forming properties of muscle proteins are one of the most important functional properties observed in processed meat products (1, 2, 3) and are responsible for the texture, water-holding, binding, and appearance of the meat products (4). As the variety of new meat products increase, the need to understand, modify and control protein gelation becomes more important (2). Currently, a trial and error approach is used by processors when making ingredient substitutions and process modifications. This approach is time consuming and expensive. When the biochemical basis of protein gelation and factors which influence the properties

0097–6156/91/0454–0243$06.00/0
© 1991 American Chemical Society

of finished gels are understood, processors will be able to
scientifically and economically manipulate protein properties to
develop desired textures and yields in processed meat products.

The biochemical basis for the gelation of skeletal muscle
proteins has been reviewed by several authors (1, 5), however, the
gelation properties of cardiac and smooth muscles have not been
characterized. The emulsification properties of cardiac and smooth
muscle proteins have been studied in emulsion systems (6), but not
in gel systems. Recent evidence indicates that emulsification properties of muscle proteins do not correlate well with function in a
meat product (7).

A wide range of intrinsic and extrinsic factors influence the
properties of muscle protein gels (8). Researchers are just
beginning to assess the importance of ingredient interactions on
heat-induced gel quality. The contribution of the salt soluble
skeletal proteins to processed meat texture has been studied extensively, however the modifying influence of the sarcoplasmic and
stroma proteins is less well understood. Non-muscle proteins used
in combination with the salt soluble proteins can also influence gel
properties. The contribution of non-muscle proteins to the finished
quality of muscle protein gels was reviewed by Foegeding and Lanier
(9). A wide range of gel textures and microstructures can be produced from multicomponent gels (10, 11, 12). The contribution of
each protein to the texture and microstructure of the gel will
depend on concentration, pH, ionic strength and heating temperature
(13), thus the properties which can be engineered into multi-
component gels are almost unlimited.

The objective of this paper is to demonstrate how the properties
of muscle protein gels can be manipulated by changes in processing
conditions and interactions between muscle and non-muscle proteins.
Results illustrate how simple modifications in pH and temperature
modify the texture and water-holding properties of muscle protein
gels. This paper will show how ingredient interactions can have
negative or positive effects on muscle protein gel properties
depending on the processing conditions and formulation selected.

Experimental Procedures

Materials. Salt soluble proteins (SSP) were extracted from chicken
breast as described by Wang (14). The SSP was suspended in the 0.6M
NaCl, 50 mM Na phosphate sample buffer and adjusted to the desired
pH. Beef semitendinosus muscle, heart, lung and spleen proteins
were extracted into high ionic strength (0.6M NaCl, 0.05M Na phosphate buffer, pH 7.0) soluble or SSP, low ionic strength (0.05M Na
phosphate buffer, pH 7.0) soluble (LIS, sarcoplasmic) and insoluble
fractions as described by Nuckles et al. (15). Whey protein concentrates (WPC) produced commercially by ultrafiltration were heated to
produce five treatments which contained from 27% to 98% soluble protein (16) as determined by the assay of Morr et al. (17). The WPC
were freeze dried for storage. All dried ingredients were hydrated
in the sample buffer 12 hr prior to use in experiments.

Gel Preparation and Evaluation. Proteins were combined in the
desired concentrations and mixed in a Polytron Homogenizer for 30 s
at a speed of 6 in an ice bath. Heat induced gels were prepared at

65, 80 or 90°C (14, 15, 16). Gel texture by failure testing was
measured on 10 mm diameter by 10 mm height cores using the Instron
(Model 4202, Canton, MA) at a crosshead speed of 10 mm/min with a 50
N compression cell. The gels were compressed between two lubricated
flat parallel plates. Apparent stress (gel strength) and strain
(deformability) at failure were calculated from the force time curve
as described by Hamann (18). Non-failure testing was performed
using the Rheometrics fluids spectrometer and parallel plate geometry as described by Wang (14). Expressible moisture was determined
as described by Jaurequi et al. (19) except the gels were centrifuged at 755 x g for 10 min. Gels were prepared for scanning electron microscopy (SEM) as described by Smith (20) and observed using
a JEOL 35 Scanning Electron Microscope (Japan Electronics) at 15 kV.
All means are the result of at least triplicate analyses.

Results and Discussion

Solubility Differences of Skeletal, Cardiac and Smooth Muscle Proteins.

Tissues from skeletal, cardiac and smooth muscle exhibit
large differences in protein content and distribution and consequently vary widely in protein functionality and bind values. The
causes for this variation have not been determined, however, the
differences in protein functionality may be due to differences in
protein content or composition. The protein fraction composition of
several beef muscles is illustrated in Table I (15). Beef semitendinosus (skeletal) muscle contained at least twice as much SSP
protein as the cardiac (heart) and smooth (lung, spleen) muscles.
Beef heart and lung contained the highest quantity of insoluble proteins, while spleen contained the highest quantity of LIS proteins.
Studies in a frankfurter model system suggested that the larger the
SSP fraction as a percentage of the total protein, the greater the
firmness, cohesiveness and water-holding capacity of the meat
product (15).

Table I. Protein fractions of beef tissues expressed as a
percentage of total protein. (Based on data
from Ref. 15 and 21)

Tissue	Low Ionic Strength Soluble (%)	Salt Soluble (%)	Insoluble (%)
Semitendinosus	21.8	44.9	28.2
Heart	30.1	21.2	48.7
Lung	36.3	9.6	54.1
Spleen	53.6	21.7	24.7

Functionality of SSP Fractions.

The lowest protein concentration of
the SSP fraction required to form a gel (defined as apparent stress
at failure less than or equal to 4.0 kPa) varied with the tissues.
The SSP fraction of beef semitendinosus muscle, heart, lung and
spleen formed gels at protein concentrations of 6.0%, 5.0%, 4.0% and
3.0%, respectively, in 0.6M NaCl, 50 mM Na phosphate, pH 7.0 when

heated to 70°C (21). The expressible moisture (33.5%) and deformability (0.50) of the gels were not significantly different at their
least concentration endpoints. In 6.0% SSP gels, strength and deformability significantly decreased and expressible moisture significantly increased in the following order: semitendinosus, heart,
lung, and spleen (Table II). Large differences in gel microstructure were observed between the tissues. Results indicated that
the gel forming properties of the SSP fractions were not equivalent
and thus functionality of muscle tissues cannot be compared solely
on the basis of the quantity of SSP in the tissue.

Table II. Texture and expressible moisture of beef salt-
soluble protein gels. (Based on data from Ref. 21)

Tissue	Apparent stress at failure (kPa)	Apparent strain at failure	Expressible moisture (%)
Semitendinosus	13.01[a]	0.81[a]	19.3[a]
Heart	9.23[b]	0.77[b]	23.3[b]
Lung	6.89[c]	0.70[c]	26.7[c]
Spleen	6.22[c]	0.66[d]	33.7[d]

[a,b,c,d]Any two means within the same column followed by different
letters were significantly different at $p < 0.05$.

Composition of SSP Fractions. Sodium dodecyl sulfate electrophoresis of the SSP fractions were performed on 12% acrylamide gels
(22) to determine differences in the protein composition of the SSP
fraction. Beef semitendinosus and heart SSP fractions contained
significantly larger percentages of myosin and actin compared to the
other two tissues (Table III) (21). The actin:myosin mole ratios
for beef semitendinosus and heart were 6.0 and 5.0, respectively.
Beef lung exhibited the lowest ratio of 3.2. The quantity of myosin
in the SSP fraction was strongly correlated with gel strength
(r=0.86), deformability (r=.80) and expressible moisture (r=-.82)
(21). These results suggest a new method for measuring the bind
constants or relative functionalities of meat ingredients used in
least cost formulation calculations by meat processors. It may be
possible to predict the gel strength and other functionalities of
all muscle tissues based on the size of the SSP fraction and amount
of myosin in that fraction. Currently, very poor correlations exist
between bind constants measured by emulsification capacity and the
actual function of meat ingredients in a formulation.

Influence of Stromal and Sarcoplasmic Proteins. Other muscle proteins can also modify the properties of SSP gels. The influence of
the sarcoplasmic or low ionic strength (LIS) proteins and the insoluble or stroma proteins on the gel properties of SSP gels from
beef skeletal, cardiac and smooth muscle tissues were determined
(21). LIS and insoluble protein fractions were substituted into SSP
protein solutions at 8.33, 16.7, 25, 33.3 and 50% to prepare gels

with a total protein content of 6.0% (w/w) in 0.6 M NaCl, 50 mM Na phosphate, pH 7.0 and heated to 70°C.

Table III. Myosin and actin composition of beef tissue salt soluble protein fraction. (Adapted from Ref. 15)

Tissue	Myosin (%)	Actin (%)	Actin:Myosin Mole Ratio
Semitendinosus	50.7	21.6	6.0[a]
Heart	47.9[b]	20.6[b]	5.0[b]
Lung	37.6[c]	10.5[c]	3.2[d]
Spleen	22.6[d]	10.1[c]	4.5[c]

[a]Source: Lowey, S. and Risby, D. 1971. Nature 234:81.
[b,c,d]
Any two means within the same column followed by different letters were significantly different at $p < 0.05$.

The addition of the LIS fraction to the SSP gels resulted in a significant decrease in gel strength of beef semitendenosus muscle (Table IV). The largest decrease in gel strength was observed at the 8.3% substitution level. Substitution of 16.7% to 41.7% LIS proteins into SSP gels did not significantly decrease gel strength in comparison to 6% (w/o) protein gels made entirely from SSP. Deformability decreased gradually as the LIS concentration in the gels increased. Expressible moisture decreased when SSP gels were substituted with 8.3% LIS proteins, but increased at higher substitution levels. LIS substitutions up to 16.7% of the SSP proteins did not have a significant deleterious effect on expressible moisture. Small quantities of LIS or sarcoplasmic proteins had an adverse effect on the strength and deformability of the SSP gel matrix. Sarcoplasmic proteins may interfere with SSP cross-linking during matrix formation as they do not form gels and have poor water-holding abilities (23). Venegas et al. (24) reported that meat homogenates containing large quantities of sarcoplasmic proteins have good water-holding capacity, but poor gel strength.

Scanning electron micrographs of SSP and SSP/LIS protein gels revealed very different microstructures (21). The SSP gels exhibited a fibrous network with distinct globular chains, while the SSP fibers in the SSP/LIS gel network appeared to be coated with a layer of LIS proteins. It is possible that a small amount of the hydrophilic sarcoplasmic proteins may enhance the water-holding capacity of SSP gels by increasing the surface area of the protein fibers in the matrix.

The addition of 8.3% insoluble proteins to the SSP gels had no significant effect on gel strength or deformability, however, higher substitutions decreased these properties (Table IV). Expressible moisture increased in all tissues as the quantity of insoluble proteins in the 6% protein gels increased. The microstructure of 6% SSP gels were different from those of SSP gels substituted with 33% insoluble protein (21). The addition of insoluble proteins appeared to interrupt the regular fibrous network causing large spaces in the gel matrix.

Table IV. Substitution of 6.0% (w/w) semitendinosus salt-soluble protein gels with low ionic strength or insoluble protein fractions. (Based on data from Ref. 21)

Substitution (%)	Apparent Stress at Failure (kPa)	Apparent Strain at Failure	Expressible Moisture (%)
Low Ionic Strength Protein Fraction			
0.0	12.8[a]	0.79[a]	19.3[a]
8.3	6.1[b]	0.80[a]	17.7[b]
16.7	4.9[c]	0.68[b]	19.5[a]
25.0	4.6[c]	0.60[c]	20.8[c]
33.3	4.4[c]	0.58[c]	25.5[d]
41.7	4.3[c]	0.54[d]	30.8[e]
50.0	3.7[d]	0.52[d]	37.1[f]
Insoluble Protein Fraction			
0.0	13.0[a]	0.81[a]	19.3[a]
8.3	12.8[a]	0.80[a]	22.7[b]
16.7	9.1[b]	0.71[b]	24.7[c]
25.0	6.7[c]	0.68[b]	28.5[d]
33.3	6.3[c]	0.64[c]	35.5[e]
41.7	5.1[d]	0.62[c]	37.5[f]
50.0	4.1[e]	0.61[c]	40.7[g]

[a, b, c, d, e, f, g]   Any two means within the same column followed by different letters were significantly different at p< 0.05.

Effect of Temperature and pH on SSP Gel Properties. Temperature and pH have a large influence on the properties of SSP gels. Wang (14) examined the dynamic rheological properties of 3.0% chicken breast SSP solutions at 4 pHs (4.5, 5.5, 6.5 and 7.5) during heating at a rate of 1°C/min in a Rheometrics fluids spectrometer at 1% strain and frequency of 10 rad/sec. Gel rigidity as determined from the complex modulus at 80°C decreased in the following order: pH 5.5 > 6.5 > 7.5 > 4.5. Expressible moisture did not follow the same pattern, as expressible moisture increased in the following order: pH 6.5 < 7.5 < 5.5 < 4.5. Gel properties were probably influenced by the protein transitions observed during heating (Figure 1). The SSP solution at pH 4.4 did not undergo any detectable rheological transitions during heating, however, several transitions were observed for SSPs at the other pHs. As pH increased, the temperature of the first transition increased from 35.5°C at pH 5.5 to 47.0°C at pH 7.5. The shape of the transitions also differed with pH indicating the influence of protein charge on gel properties.
   A range of microstructures were observed in the gels. Repre-

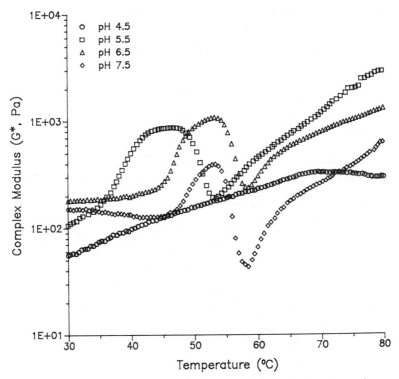

Figure 1.   Effect of pH on the complex moduli of 3% solutions of chicken breast salt-soluble proteins in 0.6 M NaCl heated at 1°C/min.

sentative scanning electron micrographs of SSP gels heated to 55°C and 65°C are presented at 10,000 X magnification in Figures 2 and 3, respectively. The gels prepared at pH 4.5 exhibit highly aggregated, globular structures with no network formation and are similar at both heating temperatures. These microstructures are consistent with the poor gel strength and high expressible moisture observed. Gels at pH 5.5 heated to 55°C were filamentous and exhibited an irregular network structure with large holes in the matrix, while regular lacy networks were observed at pH 6.5 and 7.5. At 65°C, the protein network became thicker and more regularly spaced which is indicative of increased gel strength and water-holding capacity.

These results suggest it is possible to produce a range of properties in muscle protein gels simply by controlling the pH and heating temperature. These factors are relatively easy to control and could be used by meat processors to manipulate the textures and yields in processed products.

Influence of Non-meat Proteins. Other proteins used in a meat formulation also influence the gelation properties of the SSPs and the resultant texture and yield in a processed meat product. Three types of multicomponent gels have been defined by Tolstoguzov and Braudo (1983): filled, mixed and complex. Non-meat proteins may act as fillers within the interstitial spaces of the SSP gel network. Fillers are defined as macromolecules which do not form a gel matrix themselves, but fill the interstitial spaces of a gel matrix (10).

Some authors have reported that filler materials increase gel strength and have a minimal effect on gel deformability (13, 27, 26), however, mechanical properties of filled gels depend on the properties of the matrix gel, the shape, deformability and volume fraction of the filler and the interactions between the filler and matrix (10, 11, 28). Fillers have been classified as active or inactive. Inactive fillers do not strengthen the gel matrix (11) while active fillers increase matrix strength. A mixed gel is one in which two or more different types of macromolecules form gel matrices. Multicomponent gel theory predicts that mixed and filled gels will have widely different properties and thus would expand the textural attributes of meat products which could be obtained from a formulation (10, 11, 13, 26).

Beuschel (16) investigated the gelation properties of a WPC and chicken breast SSP when heated to 65°C or 90°C in 0.6 M NaCl, 50 mM Na phosphate buffer, pH 7.0. Chicken breast SSP gels, WPC gels and combination gels were prepared by heating mixtures in 0.6 M NaCl, 50 mM Na phosphate, pH 7.0 to 65°C and 90°C. The two temperatures were selected to evaluate SSP/WPC gel properties heated below and above the denaturation temperature of the major whey protein, beta-lactoglobulin. Beta-lactoglobulin is primarily responsible for the gelation of WPC. The two major whey proteins, alpha-lactalbumin and beta-lactoglobulin denature at approximately 68°C and 80°C, respectively, although the transition temperatures are influenced by pH, ionic strength and other factors (25). Myosin in chicken breast SSP denatures at approximately 55°C. Wang (14) reported that heating SSP above 70°C did not improve gel strength.

The temperature had a large influence on the properties of the

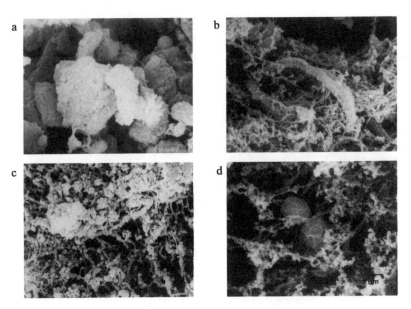

Figure 2. Microstructure of chicken breast salt-soluble protein gels in 0.6 M NaCl heated to 55°C at a) pH 4.5 b) pH 5.5 C) pH 6.5 and d) pH 7.5.

Figure 3. Microstructure of chicken breast salt-soluble protein gels in 0.6 M NaCl heated to 65°C at a) pH 4.5 b) pH 5.5 C) pH 6.5 and d) pH 7.5.

protein gels. Whey protein concentrate did not gel at $65^{\circ}C$
(Table V). Gels prepared with 4% SSP and 8% or 12% WPC protein were
only slightly stronger than gels prepared with 4% SSP alone. SSP
gels prepared with 16% WPC were weaker than those prepared with 4%
SSP probably due to the interference of WPC in the SSP protein network. SSP gels were most deformable. At $65^{\circ}C$, deformability
decreased as WPC concentration increased in SSP gels (Table VI).
With each 4% increase in WPC the deformability decreased by 0.2
units. Expressible moisture was strongly influenced by WPC concentration in the gels (Table VII). The addition of 8% and 12% WPC to
4% SSP gels caused approximately a 50% decrease in expressible
moisture at $65^{\circ}C$.

Table V. Strength of chicken salt-soluble protein (SSP)
and whey protein concentrate (WPC, 98% soluble) gels
(Based on data from Ref. 16)

Treatment	Percent Protein in SSP:WPC Gels	Apparent Stress at Failure (kPa) Temperature	
		$65^{\circ}C$	$90^{\circ}C$
SSP	4:0	24.03[b]	no gel
	8:0	76.74[a]	53.67[c]
WPC	0:8	no gel	no gel
	0:12	no gel	15.32[f]
	0:16	no gel	28.07[e]
	0:20	no gel	94.85[b]
SSP:WPC	4:8	27.78[b]	46.67[d]
	4:12	27.09[b]	97.66[b]
	4:16	15.21[c]	134.29[a]

a, b, c, d, e, f Any two means within the same column followed by
different letters were significantly different at p< 0.05.

The WPC probably functions as a particulate filler in SSP gels
at $65^{\circ}C$ as this temperature is below the denaturation temperature
for the whey proteins. Our results are consistent with filled gel
theory, where WPCs act as an inactive filler within the SSP matrix.
The addition of inactive fillers, such as wheat gluten and soy protein isolates, do not usually increase gel strength, but do increase
water-holding capacity in processed meat products (9, 29, 30, 31).
Wheat gluten proteins and soy proteins do not denature at temperatures typically used in meat processing and probably act as fillers.
The SSP gels heated to $90^{\circ}C$ were weaker than gels prepared at
$65^{\circ}C$. The WPC formed gels at $90^{\circ}C$ which increased in strength with
protein concentration. Gels prepared with SSP and WPC were stronger
than WPC gels of the same protein concentration (Table V). Gels
prepared with 4% SSP and 12% WPC (16% protein gels) were 3.5 times
stronger than 16% WPC gels. The effect decreased as protein concentration increased. Gels prepared with 4% SSP and 16% WPC were only
1.5 times stronger than 20% WPC gels. At $90^{\circ}C$, deformability was

constant in SSP/WPC gels and was dictated by the WPC (Table VI).
The SSP/WPC gels bound more water than WPC gels at the same protein
concentration (Table VII). Gels prepared with 4% SSP and 12% WPC
bound 4.3 times more water than 16% WPC gels. There appears to be
an optimum combination of SSP and WPC proteins. The 4% SSP/12% WHC

Table VI. Deformability of chicken salt-soluble protein (SSP)
and whey protein concentrate (WPC, 98% soluble) gels
(Based on data from Ref. 16)

| Treatment | Percent Protein in SSP:WPC Gels | Apparent Strain at Failure Temperature | |
		$65^{\circ}C$	$90^{\circ}C$
SSP	4:0	1.61^a	no gel
	8:0	1.61^a	1.61^a
WPC	0:8	no gel	no gel
	0:12	no gel	0.89^b
	0:16	no gel	0.69^c
	0:20	no gel	$0.88^{b,c}$
SSP:WPC	4:8	1.20^b	$0.88^{b,c}$
	4:12	$1.04^{b,c}$	$0.83^{b,c}$
	4:16	0.81^c	$0.72^{b,c}$

[a,b,c]Any two means within the same column followed by different
letters were significantly different at $p < 0.05$.

gels exhibited the largest increases in gel strength and expressible
moisture over the same concentration of WPC protein alone. A synergistic effect occurred in the gels when both SSP and WPC proteins
were heated to temperatures above their denaturation temperatures.
At $90^{\circ}C$, the SSP and WPC probably form mixed gels as both protein
fractions are denatured and reaggregate to form gels below this
temperature. The proteins formed a gel matrix which was stronger
and held more water that either protein alone. This type of result
has been reported in other mixed gel systems (32, 33).
     The degree of denaturation of non-meat protein additives is
another factor determining gel properties. Whey protein concentrates were heated to insolubilize the proteins to produce five concentrates with solubilities ranging from 27% to 98% (16). Gels were
prepared from 4% SSP/12% WPC. Interactions between the denatured
WPC proteins and SSP occurred during heating which altered the properties of the gel matrix. The use of denatured WPC improved SSP
gel strength and deformability at low temperatures compared to undenatured (98% soluble) WPC. Denatured whey proteins interacted with
SSP to enhance the gel matrix at $65^{\circ}C$. At $65^{\circ}C$, gel strength and
deformability increased as the extent of WPC insolubilization decreased to 41%. Gels prepared with 41% insolubilized protein were
over 3 times stronger than those produced with 98% soluble WPC.
Similarly, expressible moisture of SSP/WPC gels decreased as the
extent of the WPC denaturation increased. It is not clear whether
the denatured WPC were acting as active fillers within the inter-

stitial spaces of a SSP gel matrix or whether there was direct
interaction between SSP/WPC protein fibers to form a mixed or complex gel (28). The denatured WPC were probably functioning as
active fillers in the SSP gels as the denatured WPCs alone did not
produce gels when heated to 65°C. Gel properties were very different at 90°C. Gel strength decreased and deformability increased
as the extent of WPC insolubilation increased in both WPC and
WPC/SSP gels. Gel strength of 98% soluble WPC was approximately
twice that of 41% soluble WPC at 90°C. It is evident that the denatured WPCs were not able to form a mixed gel matrix with the SSPs,
as gel strength decreased to that of the SSPs alone.

Table VII. Expressible moisture of chicken salt-soluble
protein (SSP) and whey protein concentrate (WPC,
98% soluble) gels (Based on data from Ref. 16)

Treatment	Percent Protein in SSP:WPC Gels	Expressible Moisture Temperature 65°C	90°C
SSP	4:0	38.0[a]	32.3[c]
	8:0	13.1[c]	16.8[d]
WPC	0:8	no gel	63.5[a]
	0:12	no gel	41.8[b]
	0:16	no gel	31.8[c]
	0:20	no gel	20.3[d]
SSP:WPC	4:8	16.0[b,c]	17.8[d]
	4:12	17.7[b,c]	7.3[e]
	4:16	21.2[b]	5.7[e]

[a,b,c,d] Any two means within the same column followed by different
letters were significantly different at p< 0.05.

Conclusions

The properties of muscle protein gels can be manipulated by a
variety of factors and these factors could be exploited by the meat
processing industry when engineering new products. Proteins from
skeletal, smooth and cardiac muscles produce a range of rheological
properties in heated-induced gels. The functionality of these proteins in gels can be predicted based on the myosin content of the
tissues. Gel properties are modified by the amount of stroma, sarcoplasmic and non-muscle proteins in the gel system. A wide range
of properties can be engineered into gels by proper selection of pH,
processing temperature and protein ingredients. The quantity and
the extent of protein denaturation of each protein ingredient in a
SSP gel alters the rheological and water-holding properties of the
gel. Gel properties can be successfully modified if the biochemical
basis for protein gelation is understood and scientifically
exploited.

Acknowledgments

The technical assistance of Bryan Beuschel, Rod Nuckles and Shue Fung Wang is gratefully acknowledged.

Literature Cited

1. Foegeding, E.A. Food Technol., 1988, 42 (6), 58, 60-62, 64.
2. Smith, D., Food Technol., 1988, 42 (4), 116-121.
3. Whiting, R.C. Food Technol., 1988,42 (4), 104-114, 210.
4. Hermansson, A.M.; Harbitz, O.; Langton, M.J. Sci. Food Agric. 1986, 37, 69.
5. Asghar, A.; Samejima, K.; Yasui, T. CRC Crit. Rev. Food Sci. Human. Nutr., 1985, 22, 27.
6. Comer, F.W.; Dempster, S. Can Inst. Food Sci. Technol., 1981, 14, 295-302.
7. Regenstein, J.M. Proc. 41st Ann. Recip. Meat Conf., 1988, 40.
8. Smith, D. Proc. 41st Ann. Recip. Meat Conf. 1988, 48.
9. Foegeding, E.A.; Lanier, T.C. Cereal Foods World, 1987, 32, 202-205.
10. Oakenfull, D. CRC Crit. Rev. Food Sci. Nutr. 1987, 26, 1-25.
11. Morris, V.J. Chem. Ind. 1985, 159.
12. Tolstoguzov, V.B.; Braudo, E.E. J. Texture Stud., 1983, 14, 183.
13. Foegeding, E.A. Food Proteins, Amer. Oil Chem. Society: Champaign, IL, 1989; p 185-194.
14. Wang, S.F. M.S. Thesis, Michigan State University, East Lansing, MI, 1990.
15. Nuckles, R.O.; Smith, D.M.; Merkel, R.A. J. Food Sci. 1990, 55, in press.
16. Beuschel, B.C. M.S. Thesis, Michigan State University, East Lansing, MI, 1990.
17. Morr, C.V.; German, B.; Kinsella, J.E.; Regenstein, J.M.; Van Buren, J.P.; Kilara, J. A.; Lewis, B.A.; Mangino, M.E. J. Food Sci., 1985, 50, 1715-1717.
18. Hamann, D.D. In: Physical Properties of Foods; Peleg, M.; Bagley, E.B., Eds.; AVI: Westport, CT, 1983; Chapter 13.
19. Jaurequi, C.A.; Regenstein, J.M.; Baker, R.C. J. Food Sci. 1981, 46, 1271.
20. Smith, D. J. Food Sci. 1987, 52, 22-27.
21. Nuckles, R.O. Ph.D. Thesis, Michigan State Univ., East Lansing, MI, 1990.
22. Smith, D.M.; Brekke, C.J. J. Agric. Food Chem. 1985, 33, 631.
23. Acton, J.C.; Ziegler, G.R.; Burge, D.L. CRC Crit. Rev. Food Sci. Nutr., 1983, 18, 99.
24. Venegas, O; Perez, D; DeHombre, R. Proc. 34th Int. Cong. Meat Sci. Technol., 1988, 402
25. DeWit, J.N.; Klarenbeek, G. J. Dairy Sci., 1984, 67, 2701-2710.
26. Tolstoguzov, B.B.; Braudo, E.E. J. Texture Stud., 1983, 14, 183.
27. Foegeding, E.A. Proc. 41st Recip. Meat Conf., 1988, 44.
28. Ring, S.G.; Stains by, G. Prog. Food Nutr. Sci., 1982, 6, 323.
29. Comer, F.W. Can. Inst. Food Sci. Technol., 1979, 12, 157.

30. Randall, C.J.; Raymond, D.P.; Voisey, P.W. J. Inst. Can. Sci. Technol., 1976, 9, 216.
31. Sofos, J.N.; Noda, I.; Allen,C.E. J. Food Sci., 1977, 42, 879.
32. Morris, V.J.; Chilvers, G.R. J. Sci. Food Agric., 1984, 35, 1370.
33. Thom, D.; Dea, I.C.M.; Morris, E.R.; Powell, D.A. Prog. Food Nutr. Sci., 1982, 6, 97.

RECEIVED May 16, 1990

Chapter 18

Gelation of Myofibrillar Protein

E. Allen Foegeding[1], Clark J. Brekke[2], and Youling L. Xiong[3]

[1]Department of Food Science, North Carolina State University,
Raleigh, NC 27695–7624
[2]Food Science and Human Nutrition, Washington State University,
Pullman, WA 99164–6376
[3]Department of Animal Sciences, University of Kentucky,
Lexington, KY 40546–0215

Myofibrillar protein fractions from poultry white (breast) and red (thigh or leg) meat form gels with different rheological properties. Firm, deformable gels are produced with breast myofibrillar protein, while the gels formed with red muscle myofibrillar protein are lower in textural properties. The differences in gelation of myofibrillar protein at equal pH does not appear to be due to temperature of protein unfolding or protein extractability. Rather, the association process, bonding within the gel matrix and the geometry of the gel matrix are involved in the superior gelling ability of breast myofibrillar protein.

The "art" of making processed meats, which has been with us since antiquity, has produced a cornucopia of products. The quality of these products, like any food item, is judged by the desirability of color, flavor and texture. Variation in product quality is endemic to "art" based processes. Therefore, it has been a goal of food scientists to understand the physical/chemical mechanisms which produce both desirable and undesirable characteristics in meats.

Protein gelation has been used as a model system to understand the mechanisms responsible for texture and water holding, with most direct application to finely comminuted products (1, 2). In the investigation of muscle protein gelation, one has to make a choice on complexity of the system to study. Early research on the functional properties of muscle proteins established that myosin was the main protein responsible for binding together sausage structure (3, 4). Therefore, myosin gelation has been the subject of many investigations; for recent reviews see Asghar et al. (2) and Ziegler and Acton (5). A myosin suspension would represent the simplest system for investigating functionality of meat proteins. At the other extreme, a meat product such as a frankfurter is the most complex in that it contains numerous different muscle proteins and other ingredients that would affect gelation (6). In order to develop a fundamental understanding of the chemical/physical events that occur during the manufacturing of processed meat products, investigations on myosin and frankfurters, along with intermediate systems, are required.

Salt soluble protein suspensions, composed of extracted (ca. 0.5-0.6 ionic strength) myofibrillar proteins, and myofibril suspensions, composed of isolated myofibrils (ca. 0.1-0.3 ionic strength), are preparations which are intermediate between myosin and meat that have been used to investigate the functionality of meat proteins. The main difference between these preparations and meat is that the connective tissue, lipids and cell membranes have been removed. The myofibril contains the contractile system of skeletal muscle. Proteins which participate in contraction (actin and myosin), regulators of contraction (troponins and tropomyosin) and structural proteins (alpha-actinin, C-protein, M-protein, nebulin, titin) are found within the sarcomere organized in ordered biological structures: the thick filaments, thin filaments, Z-line and M-line (7). The function of salt and chopping in processed meat manufacturing is to extract the myofibrillar proteins from their biological structures. Myofibrils are similar to muscle in that protein extraction is related to functionality of both.

Salt soluble protein (SSP) is used to describe proteins extracted from muscle by saline solutions. Unlike myofibrils, there is no standard for this preparation so the proteins comprising SSP will depend on the isolation method. In comparison to myofibrils, SSP represents the proteins extracted and soluble under defined conditions.

Factors which are associated with texture and water holding of muscle protein gels are: pH (8, 9, 10, 11), ionic strength (8, 9, 10, 12), protein extractability (13), protein solubility (8, 9), protein concentration (10), myosin:actomyosin (actin) ratio (14, 15, 16, 17) and protein isoforms (18, 19, 20). This work will discuss the role of these factors in the gelation of myofibrils and salt soluble protein.

Gelation of Myofibrils and Salt Soluble Protein

Protein Concentration. That protein concentration would have an effect on properties of thermally-induced protein gels is obvious and well-documented. However, proteins from different muscle systems may differ in their reaction to equal changes in protein concentration. For example, in Figure 1, the effect of concentration on gel penetration stress for chicken breast and leg myofibril suspensions is shown. It is apparent from the data that there is a definite concentration effect for both breast (white) and leg (red) proteins. In addition, it is also obvious that there are differences in the slopes of the two curves between red and white myofibril gels. Similar protein concentration effects on gel shear stress at failure (strength) are seen with turkey SSP (8) and myosin/actomyosin (14) (Table I). Shear strain at failure (deformability) also increases with protein concentration, although to a much smaller extent (Table I). These results indicate that there are definite differences between gelation properties of red and white myofibrils and SSP. By looking further at some of the factors which contribute to this difference, it should be possible to postulate and define factors important to the basis of SSP gelation.

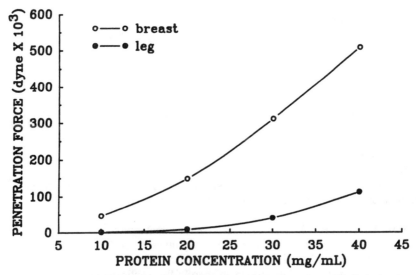

Figure 1   Effect of protein concentration on gel strength of myofibrils isolated
                  from chicken breast and leg muscles. Data points represent means of
                  three replicates.   Gelation conditions: 0.6 M NaCl, 50 mM
                  piperazine-N, N bis (2-ethane sulfonic acid) (PIPES), 1 mM NaN$_3$,
                  pH 6.00; heating rate = 1°C/min, from 18 to 70°C
                  (Reprinted with permission from ref. 23. Copyright 1989.)

Table I. Rheological Properties of Turkey Breast and Thigh
Myofibrillar Protein Gels

PROTEIN	Protein conc. (mg/ml)	pH	Stress (kPa)	Strain (m/m)
Breast				
Salt Soluble Protein[1]	25	6.0	3.75[a3]	2.11[a]
	35	6.0	7.48[b]	2.22[b]
Myosin/Actomyosin[2]	10	7.0	2.60[x]	1.96[xy]
	15	7.0	5.14[z]	2.06[y]
Thigh				
Salt Soluble Protein	40	6.0	8.37[b]	2.07[a]
Myosin/Actomyosin	20	7.0	3.54[y]	1.75[x]
	25	7.0	5.03[z]	2.12[y]

[1]Data from (8).
[2]Data from (14).
[3]Values for each rheological property within one type of myofibrillar protein with different superscripts are significantly different ($P < 0.05$).

Myosin/Actomyosin (Actin). The role of different muscle protein fractions in thermally-induced gelation has been extensively investigated (2, 15, 16, 17, 19, 21). Some of the points that have been established from these investigations include: myosin is the major gelling protein in myofibrils; actin alone has poor gelling properties but shows a synergistic effect to myosin gelation; and actomyosin has excellent gelling ability. Thus, gel rheology of a protein system containing myosin, actin and/or actomyosin is dependent on the relative amount of each of the protein components in the mix. For example, gel rigidity of reconstituted actomyosin increases proportionally with the myosin/actomyosin ratio; the maximum in gel rigidity is generally seen at a ratio of around four (16, 17, 19). Therefore, for meaningful comparisons between SSP fractions, it is necessary to know the protein composition of the SSP under study. One of the factors that has been found to influence the ratio of myosin to actin is extraction time (Table II), with the ratio increasing with increasing extraction times. For both unextracted myofibrils and up to 24 hr extraction, myosin:actin was greater for leg SSP, being 3.1 and 2.3 for leg and breast, respectively, at 24 hr. extraction (Table II).

pH. Electrostatic or ionic interactions are considered to be one of the major forces involved in gel formation (2). This type of interaction is pH dependent. Optimal pH for gelation of myofibrillar preparations is generally around 6.0 under normal conditions of meat processing that require 2.5 - 3.0% NaC1. However, different pH

Table II. Myosin to Actin Ratio[1] in the Extracts of Chicken
Myofibril Suspensions Stored at 4°C for Various Times[2]

	Fresh Myofibrils	11 min extract	2 hr extract	24 hr extract
Breast	2.1	1.5	2.0	2.3
	(0.2)	(0.1)	(0.2)	(0.3)
Leg	2.4	2.8	3.0	3.1
	(0.2)	(0.1)	(0.1)	(0.2)

[1]Ratio of myosin to actin on a weight basis. Data and means from three replications with standard deviations in parentheses.
[2]Data from (22).

values are observed for optimum gelation of specific myofibrillar components and protein isoforms for different animal species and muscle types. Although work by Morita et al. (19) with chicken indicated that breast and leg myosins have different pH optima for rigidity development, pH 5.5 and 5.1 for respective breast and leg samples, we have found that turkey breast and thigh SSP forms gels with maximum rigidity at pH 6.0 in 0.5 M NaCl (8).

Protein Association and Thermal Transitions. Dudziak et al. (14) showed that myosin and actomyosin from turkey breast and thigh have very similar thermal transitions during the gelation process. However, the gels formed from the breast and thigh proteins vary greatly in rheological properties, suggesting that some variation in the aggregation process is responsible for producing the differences in rheological properties. There are various techniques used to determine protein unfolding and association during thermal denaturation, including differential scanning calorimetry (DSC), hydrophobicity change (as indicated by a fluorescent probe) and turbidity changes. Transition temperatures (designated as Tm) are used to identify midpoints where conformational changes in proteins occur upon the absorption of thermal energy. It is well established that gel formation results from protein-protein interactions that produce a three-dimensional, cross-linked, continuous gel matrix upon heating (13). The sequence of the events, including unfolding (denaturation), loss of solubility (aggregation) and subsequent gelation, can play a major role in determining the characteristics of the gel.

The increase in hydrophobicity (as measured by a fluorescent probe) at pH 6.0 that results from increasing thermal input is shown in Figure 2. Although the relative fluorescence was initially greater for leg SSP, leg and breast generally showed similar changes in hydrophobicity (fluorescence curve shape) during the heating process. Likewise for DSC, both breast and leg pH 6.0 SSP preparations show a single endothermic peak at the same transition temperature (55-60°C), and the shape of the curves also shows no pronounced differences between breast and leg SSP (Xiong, Y.L.; Brekke, C.J. *J. Food Sci.*, in press). Similar DSC and hydrophobicity transitions during gel formation were also shown for myosin/actomyosin from turkey breast and thigh at pH 7.0 (14).

However, onset, rate and extent of protein-protein interaction, as indicated by turbidity change, differs dramatically for breast and leg (Figure 3), with a faster rate and greater extent for leg SSP. Another indication of differences in aggregation between breast and leg SSP are shown by Arrhenius plots of the SSP interaction at maximum rate of aggregation (Xiong, Y.L.; Brekke, C.J. J. Food Sci., in press). Both breast and leg reactions follow first order kinetics, and activation energies, derived from the slopes of the Arrhenius plots, provide further conformation that leg SSP requires more heat input and is more temperature dependent than breast SSP during thermal aggregation (Table III).

The approximate onset temperatures for thermally-induced protein unfolding, protein-protein aggregation and subsequent gel formation, are given in Table III. Although thermally-induced changes in protein unfolding, interactions and gelations all occur over a temperature range, and there are temperature overlaps (Xiong, Y.L.; Brekke, C.J. J. Food Sci., in press), several temperature differences in Table II can be noted. There is a small difference for protein unfolding onset; however, the onset of protein aggregation, as incicated by turbidity, occurs at a higher temperature for leg SSP. Onset of gelation, on the other hand, occurs at the same temperature for both breast and leg SSP. Perhaps more important than the fact that aggregation for breast occurs at a lower temperature than leg SSP is the fact that the magnitude of the difference in onset temperatures between protein unfolding and aggregation and between aggregation and gelation differs for breast and leg. The 4°C difference between protein unfolding and aggregation for breast SSP, versus a 10°C difference with leg SSP, seems to be the opposite of what it should be to explain the greater strength of breast SSP gels compared to leg SSP gels. Perhaps more important is the fact that there is a grater difference for breast between

Table III. Onset Temperature (°C) for Thermally Induced Protein Unfolding, Protein-Protein Aggregation and Gel Formation, and Apparent Activation Energy (E_a) for Protein-Protein Aggregation of Chicken Salt-Soluble Proteins

	Protein Unfolding (ANS-fluorescence) (°C)	Aggregation (turbidity) (°C)	Gelation (°C)	E_a (J/mol)
Breast	32	36	48	244
Leg	30	40	48	718

Data from (23).

temperature of aggregation and gelation than there is for leg SSP. Thus, for any rate of temperature increase during gel formation, breast SSP would have a longer period of time to aggregate prior to onset of gelation. That is, this lag between

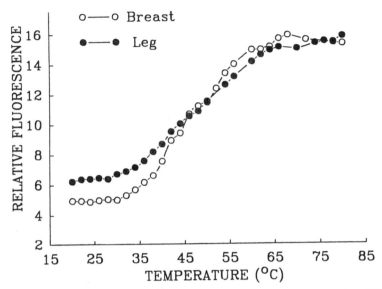

Figure 2   ANS-protein fluorescence intensity of salt-soluble proteins (SSP)
prepared from postrigor chicken myofibrils as a function of
temperature. Data points represent means from three replicates. SSP
was suspended (1 mg/mL) in 0.6 M NaCl, 50 mM PIPES, 1 mM
NaN$_3$, pH 6.00; heating rate = 1°C/min
(Reprinted with permission from ref. 23. Copyright 1989.)

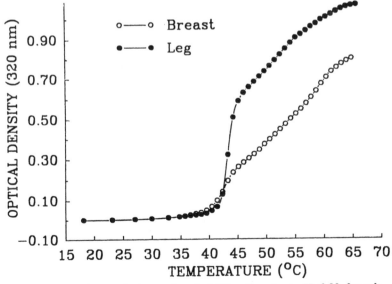

Figure 3   Protein-protein interaction (turbidity change) at pH 6.00 for salt-
soluble proteins extracted from postrigor chicken breast and leg
myofibrils. Assay conditions: 0.3 mg/mL protein, 0.6 M NaCl, 50
mM PIPES, 1 mM NaN$_3$, pH 6.00; heating rate = 1°C/min
(Reprinted with permission from ref. 23. Copyright 1989.)

initial aggregation and gelation may allow formation of a gel microstructure that is more conducive to gel strength.

First derivative plots can also be used to examine in more detail the differences in protein-protein interaction (turbidity change). The plot for both breast and leg at pH 6.0 is shown in Figure 4. There is a dramatic difference in peak heights (5-6 times greater for leg than for breast), indicating a difference in rate of protein-protein interaction. Tm1 and Tm2 at 43 and 53°C, respectively, are the same for breast and leg. However, breast proteins showed a third peak at about 60°C. This suggests that different protein-protein interaction mechanisms are involved for red vs. white muscle proteins.

Protein Extraction

From the preceding section, it is apparent that there are some definite differences in the rheological properties of gels made from white (chicken or turkey breast) vs. red (chicken leg or turkey thigh) SSP. However, for SSP to have an effect on gelation and, subsequently, on quality characteristics of processed meat products, it must first be extracted from the myofibril. This section will address how extraction from the myofibrillar structure fits into the overall gelation mechanism.

Maximum penetration force for breast and leg SSP is at about 280 and 175 x 10^3 dynes, respectively, after 20 mg/ml suspensions are heated to 80°C (Xiong, Y.L., Brekke, C.J. J. Food Sci., in press). Under the same conditions, including total protein content, the maximum penetration force for myofibrils from breast and leg is only approximately 160 and 10 x 10^3 dynes for breast and leg, respectively (Xiong, Y.L., Brekke, C.J. J. Food Sci., in press). The myofibril preparation at 0.6 M NaCl is a mixture of both extracted myofibrillar protein and non-extracted myofibrils, a condition that also would be present in a processed meat system such as a frankfurter. This fact that not all protein has been extracted obviously makes a large difference in gel strength of SSP and myofibril suspensions. Hence, extraction from the myofibril is necessary for maximum gelation of muscle proteins.

Extracted Protein Concentration. Extraction time is important to the quantity of protein extracted from the myofibrils, with a maximum for both breast and red myofibrils at approximately 11 hr storage (Xiong, Y.L., Brekke, C.J. J. Food Sci., in press). The effects of this extraction time on gel strength of the myofibril suspension (including extracted and non-extracted proteins) further supports the correlation between protein extractability and gel strength. The enhanced gel strength with the increased amount of extracted SSP is expected because at a greater concentration of SSP more protein cross-links can be formed. However, whereas extractability plateaus at approximately 11 hr storage, the rapid increase in gel strength plateaus earlier, at about 5 hr storage, indicating that extractability is not the sole explanation for variation in myofibril gel strength due to extraction time (22).

pH. Both breast and leg myofibrils exhibit a pH-dependent extraction profile, with extractability being least at pH 5.50 for either breast or leg samples (Figure 5). This

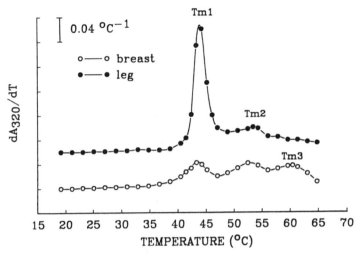

Figure 4   Derivative curves of the protein-protein interaction at pH 6.00 for salt-
soluble proteins of postrigor chicken breast and leg myofibrils.
dA_{320}/dT = differential change in optical density as a function of
temperature (Reprinted with permission from ref. 23. Copyright 1989.)

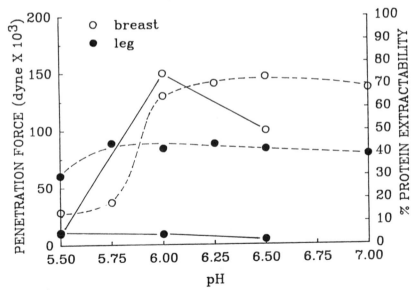

Figure 5   Effect of pH on protein extractability and gel strength of myofibrils
isolated from postrigor chicken breast and leg (mean of 3 replicates).
Extractability, - - - - - ; penetration force, ————————.
Conditions: extractability - 5 mg/mL protein, 0.6 M NaCl, 50 mM
PIPES, 1 mM NaN_3; gelation - 20 mg/mL, 0.6 M NaCl, 50 mM
PIPES, 1 mM NaN_3, heating from 20 to 70°C at 1°C/min
(Reprinted with permission from ref. 23. Copyright 1989.)

would be expected since myofibrillar proteins collectively have an isoelectric point near pH 5.1 to 5.4.   Breast myofibrils show a great dependence of protein extractability on pH between 5.75 and 6.00; at pH greater than 6.00, further change in protein extractability appears to be minor.   This suggests an increase in electrostatic repulsion among protein molecules and an increase in protein-solvent interaction as the pH was increased from 5.75 to 6.00. Extraction of proteins from leg myofibrils is less dependent on pH than for breast myofibrils, although from pH 5.50 to 5.75 there is an increased extractability of the leg myofibrillar proteins. Since the pH of chicken leg muscle is generally greater than 6.0, the minimal extractability at reduced pH is not a practical problem in processed chicken products from leg meat.

Based on extractability alone, and all other things being equal, it would be expected that the maximum gel strength of breast myofibrils would occur at pH 6.00 or greater with a drastic reduction at pH 5.75. Gel strength for leg myofibrils should not change from pH 5.75 to greater pH values. Some of this seems to be the case, as shown in Figure 5, where pH is seen to have a marked effect on gelation. Obviously, breast myofibril gelation is more pH-dependent than is leg myofibril gelation. As at least partially explained by protein extractability, minimum gel strength for those pH values investigated is obtained with breast myofibrils at pH 5.50, with a maximum at pH 6.00. However, extractability does not explain the additional decrease in breast myofibril strength at pH 6.50. This suggests that other factors, e.g. ionic or electrostatic interactions, are also involved in gel strength. Foegeding (8) also reported that turkey breast SSP (0.5 M NaCl) forms a more rigid gel at pH 6.0 than at 5.0 or 7.0. The gel strength of leg myofibril suspensions does not change from pH 5.50 to 6.50, even though protein extractability increases somewhat between 5.50 and 5.75 (Figure 5).

SDS-PAGE shows that the composition of the extracted protein (SSP) is influenced by the pH of the extraction buffer, and this pH effect is dependent upon the muscle type (Xiong, Y.L.; Brekke, C.J. J. Food Sci., in press). Actin is the predominant component of the SSP extracted from the breast myofibrils at pH 5.50, with myosin present in a small quantity compared to actin. Proteins extracted at pH 6.00 are similar to those extracted at pH 6.5°. These results are very similar to those reported by Foegeding (8) on postrigor turkey breast SSP, where differences were seen between SSP soluble at pH 5.0 and 6.0, but not between pH 6.0 and 7.0. The pH effect on protein composition is much less for leg myofibril samples, and the relative amounts of the individual proteins appears similar for samples extracted at all three pH values. Thus, both extractability and composition of extracted proteins can be influenced by pH of the environment and impact gelation mechanisms that lead to the differences observed in gel strength.

Summary

Factors in addition to protein extractability must play a role in the observed differences for gelation properties of red vs. white muscle protein systems. Since chicken SSP gelation follows the sequential changes of protein unfolding to association to gelation, the degree of protein unfolding prior to aggregation and the extent of protein aggregation before gelation (i.e., the amount of aggregated protein required for the matrix at gel point) are likely also involved. One must conclude that differences in protein gelation of these two muscle systems is dependent upon differences in mechanisms of protein-protein interaction for each.

Literature Cited

1. Acton, J. C.; Ziegler, G. R.; Burge, D. L., Jr. CRC Critical Reviews in Food Sci. and Nutr., 1983, 18, 99-121.
2. Asghar, A.; Samejima, K.; Yasui, T. CRC Critical Reviews in Food Sci. and Nutri., 1985, 22, 27-107.
3. Fukazawa, T.; Hashimoto, Y.; Yasui, T. J. Food Sci., 1961, 26, 331-336.
4. Fukazawa, T.; Hashimoto, Y.; Yasui, T. J. Food Sci., 1961, 26, 550-555.
5. Ziegler, G. R.; Acton, J. C. Food Technol., 1984, 38(5), 77-80-82.
6. Foegeding, E. A.; Lanier, T. C. Cereal Foods World, 1987, 32, 202-205.
7. Bechtel, P.J. In Muscle as Food; Bechtel, P.J., Ed.; Academic: Orlando, FL, 1986; Chapter 1.
8. Foegeding, E. A. J. Food Sci., 1987, 52, 1495-1499.
9. Ishioroshi, M.; Samejima, K.; Yasui, T. J. Food Sci., 1979, 44, 1280-1284.
10. Samejima, K.; Oka, Y.; Yamamoto, K.; Asghar, A.; Yasui, T. Agric. Biol. Chem., 1986, 50, 2101-2110.
11. Yasui, T.; Ishioroshi, M.; Nakano, H.; Samejima, K. J. Food Sci., 1979, 44, 1201-1204, 1211.
12. Hermansson, A-M.; Harbitz O.; Langton, M. J. Sci. Food Agric., 1986, 37, 69-84.
13. Samejima, K.; Egelandsdal, B.; Fretheim, K. J. Food Sci., 1985, 50, 1540-1543, 1555.
14. Dudziak, J. A.; Foegeding, E. A.; Knopp, J. A. J. Food Sci., 1988, 53, 1278-1281, 1332.
15. Ishioroshi, M.; Samejima, K.; Arie, Y.; Yasui, T. Agric. Biol. Chem., 1980, 44, 2185-2194.
16. Yasui, T.; Ishioroshi, M.; Samejima, K. J. Food. Biochem., 1980, 4, 61-78.
17. Yasui, T.; Ishioroshi, M.; Samejima, K. Agric. Biol. Chem., 1982, 46, 1049-1059.
18. Asghar, A.; Morita, J.-I.; Samejima, K.; Yasui, T. Agric. Biol. Chem., 1984, 48, 2217-2224.
19. Morita, J.-I.; Choe, I.-S.; Yamamoto, K.; Samejima, K.; Yasui, T. Agric. Biol. Chem., 1987, 51, 2895-2900.
20. Wicker, L.; Lanier, T. C.; Hamann, D. D.; Akahane, T. J. Food Sci., 1986, 51, 1540-1543, 1562.
21. Samejima, K.; Ishioroshi, M.; Yasui, T. Agric. Biol. Chem., 1982, 46, 535-540.
22. Xiong, Y. L.; Brekke, C. J. J. Food Sci., 1989, 54, 1141-1146.
23. Xiong, Y. Ph.D. Thesis, Washington State University, Pullman, WA, 1989.

RECEIVED June 20, 1990

Chapter 19

Interactions of Muscle and Nonmuscle Proteins Affecting Heat-Set Gel Rheology

Tyre C. Lanier

Food Science Department, North Carolina State University,
Raleigh, NC 27695–7624

Gel matrix development by myofibrillar (MF) proteins can be directly
influenced by chemical interactions between the non-muscle (NM)
proteins and the MF proteins, as well as indirectly by changes in the
molecular environment (contribution to total protein concentration,
water state and availability, ionic strength and types, pH) brought on
by the presence of NM proteins. Such would be expected to impact
the rheological properties of the finished food, but the presence of NM
proteins may also affect food texture independently of any effects on
the gel matrix formation by MF proteins.

The elastic structure of those processed meat, poultry, and fish products which
originate as highly comminuted pastes of muscle, water and other ingredients is
attributed to gelation of the myofibrillar proteins by heat. The texture of these
products thus depends on the structure of the matrix formed, the amounts and types
of particles and solutes entrapped by the gel matrix, and the moisture content of the
finished product (1). Each of these factors may be influenced by the presence of
added non-muscle (NM) proteins.

Formation of the Gel Matrix

Formation of the gel matrix structure by the myofibrillar proteins is a complex event
that is generally understood to be affected by the following parameters:

1. <u>Myofibrillar Protein Concentration</u>. A minimum concentration of myofibrillar
   protein is necessary for gelation to occur; below this critical concentration
   heating will effect only coagulation and precipitation. Above this
   concentration, the gel will become stronger and more rigid as the concentration
   of myofibrillar protein is increased, until water, which is required for protein
   hydration and solvation, becomes limited.

2. <u>Quantity and State of Water</u>. The charged myofibrillar proteins are soluble in
   water in the presence of salt (see subsequent discussion), being stabilized in a
   particular three-dimensional conformation by the balance between
   intramolecular forces and surface interactions with water. The polar nature of
   the solvent, water, favors clustering of hydrophobic residues within the folded

0097–6156/91/0454–0268$06.00/0

polypeptide chain to minimize decrease in entropy of the system which would result from their exposure to water at the surface. This contributes to conformational stability of protein molecules prior to heating, but may become the basis for intermolecular bonding when hydrophobic sites on adjacent protein molecules are exposed to the surface during heating. Thus water serves to initially disperse the myofibrillar protein molecules, allowing a more expanded network to develop as protein-protein bonds form during heating. When ingredients other than myofibrillar protein and salts are included in the system, water content of the system may not reflect the quantity of water available to hydrate and suspend the myofibrillar protein molecules, since these added ingredients may also tie up a portion of the solvent by chemical or physical means.

3. Ionic Type and Strength. The surface charge of proteins will depend on the amounts and types of ions present. With addition of sodium chloride, the net negative charge of myofibrillar proteins is effectively increased. This results in protein-protein repulsion, thus also contributing to dispersion of protein molecules and formation of an expanded network upon heating. Depending upon their position in the lyotropic series, ions affect the polarity of the solvent, and thus favor or inhibit the exposure of hydrophobic residues buried within the folded, solubilized proteins (2). Salting alone initiates gel formation of some fish muscle pastes (3), presumably due to its inducing both thermal destabilization of the native structure and the exposure of binding sites at the surface. Divalent cations may form cross-bridges between adjacent protein molecules, contributing to coagulation and/or gelation.

4. Time/Temperature. If binding sites of any type are exposed at the surface of suspended proteins, given sufficient time, protein concentration, and translational energy protein-protein bonds will form and a gel matrix may be gradually built up. Increasing temperature destabilizes internal hydrogen bonds, resulting in unfolding of proteins with concomitant exposure of more reactive bonding sites at the surface, and in increased translational energy, both factors contributing to an enhanced rate of intermolecular bonding.

5. pH. Surface charge of protein molecules is affected not only by the presence of dissolved salts but solvent pH as well. Charge repulsion by protein molecules is enhanced as pH is increased above the isoelectric point. As has been stated, dispersion of protein molecules is essential to the formation of an expanded network which can entrap water and other constituents of the gelling mixture. Generally, as charge repulsion increases, the minimum protein concentration and temperature for gelation to occur increases, and can reach a point at which gelation is altogether inhibited. A more dispersed and even gel network will generally be more cohesive, bind water more effectively, and scatter less light (appear more translucent) than one characterized by a lesser number of large superjunctions (Niwa, E. In Surimi Technology, Dekkar: New York, in press).

6. Interactions with Other Components. Besides water and salts, other constituents of the product formulation may, through direct or indirect interaction with the myofibrillar proteins, affect the type of gel structure formed and consequently the product texture. In some red meat and poultry products, fats are a major constituent. Protein coating of divided fat particles, often referred to as "emulsification," is initially important in attaining an even distribution of fat particles throughout the product batter. Upon heating, fat is

encased in the gel matrix formed by the myofibrillar proteins and thus stabilized against migration within and from the product. There is some evidence that direct (chemical) interactions may occur between carbohydrate hydrocolloids added to a meat batter, particularly certain gums such as carrageenan or alginate, and myofibrillar proteins, which might affect gelation of the latter. However, the greater effects of starch addition arise from its competition with protein for available water and the associated concentrating effect on the protein phase, as well as from a "filler effect" due to its occupying a specific volume fraction of the interstitial spaces within the gel matrix (4). These types of effects on gel matrix structure development and rheology also pertain to the effects of added NM proteins. Common examples of this ingredient group are wheat, egg, milk, soy, and blood proteins, most of which are dried prior to use and some of which may be fractionated (partially purified) from the original source, such as whey or soy protein isolates. While gelatin is seldom intentionally added to such processed meat product formulations, it may be present as a result of the thermal degradation of collagen naturally present in the meat source.

Gel matrix development by MF proteins can be directly influenced by chemical interactions between the NM proteins and MF proteins, as well as indirectly by changes in the molecular environment of the gelling MF proteins (total protein concentration, water state and availability, ionic strength and types, pH, etc.) brought on by the presence of NM proteins. These would be expected to impact the rheological properties of the finished food. However, the presence of NM proteins may also affect the food texture independently of their effects on gel matrix formation by MF proteins.

Models for Partitioning of NM Protein in a MF Protein Food

Ziegler (Ziegler, G; Foegeding, E. A. In Advances in Food and Nutrition Research, Academic Press, in press.) envisioned five possible models for the spatial partitioning of a gelling protein and a gelling or non-gelling coingredient (Figure 1). Consideration of these in light of low-strain rheological data collected in this laboratory may be helpful in examining the possible interactions of MF and NM proteins in a gelled food system where MF protein is the primary gelling component.

The first two of these models (Figures. 1A,B) are termed filled gels, and are distinguished depending upon the phase state of the system. In the first, the filler remains soluble in the interstitial fluid of the gel matrix, while in the second the filler exists as dispersed particles, presumably unassociated with the gel matrix. These particles may be hydrated or unhydrated, and may even consist of a secondary gel network as pictured. Tolstoguzov (5) has labeled these as type I and type II filled gels, respectively. (Figures 1C,D,E will be discussed later in the text.)

If the filler ingredient of a MF protein gel does not directly or indirectly influence the formation of the gel matrix by MF protein, the primary effects on gel rheology most likely arise from the uptake of water by the filler and its volume fraction displacement in the gel. Hydration of the filler component, which is assumed in the type I filled gel model since the filler is soluble, should add to the viscosity of the interstitial fluid and hence to the elastic modulus (rigidity) of the gel. Uptake of water by a particulate component (type II filled gel) may result in the swelling of these particles to nearly completely fill the interstitial spaces of the gel matrix. In such a case, the rigidity of the gel would be markedly increased.

The following examples illustrate these points. A type I filled gel is exemplified by a surimi-whey protein concentrate (WPC) mixture (such as might occur in a crab analog product formulation) in which the MF proteins have been "set"

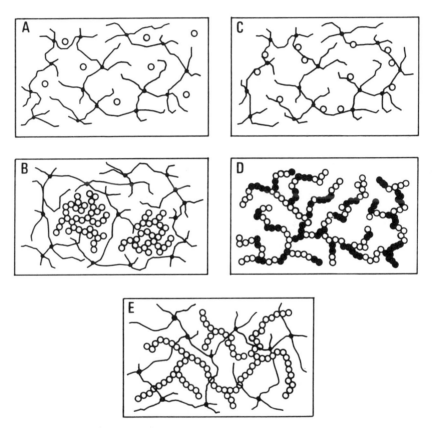

Figure 1 Possible models for the spatial partitioning of a gelling protein and a gelling or non-gelling coingredient.
(Reproduced with permission from Ref. 21. Academic Press, in press.)

to form a gel at low temperature below the denaturation/gelation temperature of the WPC. Figure 2 (6) presents a thermal scanning plot showing low-strain rigidity development in four protein systems: a surimi sol (upper curve), a whey protein solution of the same protein concentration as the surimi sol (lower curve), a 4:1 (80:20) surimi-whey protein mixture (upper middle curve) and a "control" for this mixture in which the lesser protein ingredient (whey) is replaced by water (lower middle curve). At the mid-range of temperatures (40°-60°C) the myofibrillar component has formed a rigid gel (the trough in the curve is characteristic of fish protein and does not reflect disintegration of the gel), yet the WPC has not yet left the soluble state. At this point the system matches the model of a type I filled gel. The presence of the NM protein in the system increases its rigidity despite the soluble state of the filler protein, as seen by comparing the curves of the mixture and its "control." Likely the soluble NM protein acts mainly to increase the interstitial fluid viscosity, imparting additional rigidity to the myofibrillar protein gel.

An example of a type II filled MP protein gel is Japanese kamaboko, constituted by a surimi gel matrix in which hydrated, swollen starch granules are distributed. Figure 3 (4) again shows a thermal scanning plot of low-strain rigidity, this time for a surimi sol with and without added potato starch. It is clear that upon gelatinization and swelling of the imbedded starch granules near 60°C, the overall gel rigidity is increased. Addition of certain insoluble NM proteins, though they may not be granular in nature like starch, may similarly affect the rigidity or firmness of MF protein gels.

The structural models presented in Figures 1-C and -D are what Ziegler termed "complex gels." In these the coingredient associates directly via non-specific interactions with the primary gelling component. The coingredient, whether it be non-gelling, present at non-gelling levels, or even able to gel in pure solution at the concentrations present, may associate with the gelling fraction without constituting a continuous linkage in the gel matrix (Figure 1C). Such associations may or may not affect the gelation properties of the gelling fraction, and if an effect is produced it may be positive or negative with respect to both the structural geometry of the gel matrix formed and/or its composite rheology. The non-gelling sarcoplasmic (SP) proteins of fish muscle are thought to associate with the MF proteins in just such a way to reduce their gel-forming potential (7,8) this being a primary reason why the SP proteins are removed during the process of surimi manufacture. Similarly, and in contrast to the effect of WPC addition in Figure 2, addition of EW at a 1:4 ratio with surimi of the same protein concentration seems to lower the rigidity modulus of the gel formed initially by the surimi MF proteins (Figure 4 (6)), perhaps through association with the MF proteins.

Alternatively, two or more proteins may copolymerize to form a single, heterogeneous network (Figure 1D). Such an occurrence is probably very rare, given that not only must the differing proteins possess chemically compatible binding sites, but they must also expose these under the same conditions of molecular environment and temperature. Indeed, recent studies in our laboratory of the gel-forming properties of mixtures of comminuted beef muscle with fish surimi, representing isoforms of MF proteins which one could assume are chemically quite similar, reflect differing optimum conditions of temperature and pH for gelation, and therefore a non-cooperativity in gelation. Additionally, where the microenvironmental conditions for optimal gelation of two proteins may appear to be nearly the same, differences in chemical structure may still preclude compatibility in gel network formation. A recent interesting example of this is illustrated in the failure test data of Figure 5 (T. Lanier, unpublished data). Two functionally different commercial WPC preparations (from the same manufacturer) demonstrate reasonable compatibility in gelation, as might not be totally unexpected despite the likely variability in their contents of the various whey proteins, salts, and differing structural alterations due to differing

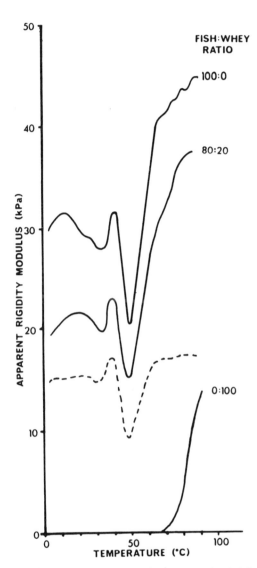

Figure 2   Thermal scanning plot showing low-strain rigidity development in four protein systems: surimi sol (upper curve), whey protein solution of the same protein concentration as the surimi sol (lower curve), 4:1 (80:20) surimi-whey protein mixture (upper middle curve) and a "control" for this mixture in which the lesser protein ingredient (whey) is replaced by water (lower middle curve).

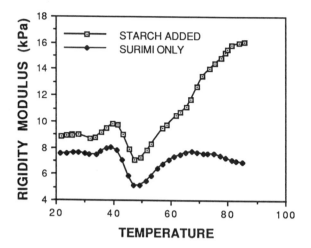

Figure 3   Thermal scanning plot showing low-strain rigidity
development in a surimi sol, with and without added potato starch.
(Reproduced with permission from Ref. 4. Copyright 1985 Institute of Food
Technologists.)

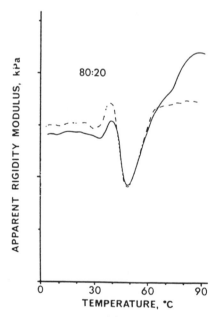

Figure 4   Thermal scanning plot showing low-strain rigidity
development in a 4:1 (80:20) surimi-egg white protein mixture (solid line)
and a "control" for this mixture in which the lesser protein ingredient (egg
white) is replaced by water (dashed line).
(Reproduced with permission from Ref. 6. Copyright 1985 Institute of Food
Technologists.)

manufacturing processes. When mixed with egg white (EW) under conditions that favor gelation of both components of the mixtures, a clear antagonism develops between one WPC and EW while a reasonable compatibility is exhibited with the other.

Nonetheless, it is thought that bovine serum albumin (BSA)-ovalbumin gels are complex, copolymerized gels (9) and such interactions have been postulated to occur between the MF protein of beef or fish and such NM proteins as blood plasma and soy (10-12). In a totally comminuted meat system, copolymerization of NM and MF proteins need not be required to produce a desirable gel structure when NM proteins are added. The rheological properties of the food may be desirably affected via an interaction of the two proteins through any of the other three mechanistic models previously discussed (Figures 1A-C).

Additionally, the spatial partitioning model of Figure 1-E might prove equally acceptable in such a case. This represents an interpenetrating network of two proteins, neither directly interacting with the other, but structurally cooperative due to the entwining of the two protein gel networks. Such an interaction between two differing proteins, both of which are able to gel independently, may be more common than the complex copolymerization of Figure 1-D, simply due to the fact that differing proteins, particularly MF as compared with NM proteins, generally are thought to gel by different mechanisms.

The marked increase in rigidity of the mixed protein gel in Figure 2 which occurred upon gelation of the WPC above 60°C logically could be explained as resulting from either formation of a type II filled gel (Figure 1-B), or an interpenetrating gel network (Figure 1-E). A copolymerized gel, at least as depicted in Figure 1-D, obviously was not formed since the surimi gelled prior to the WPC. However, this does not rule out the possibility that direct chemical interactions may have occurred between the gelled networks of the two proteins, whether interpenetrating or not, which further strengthened the composite gel. Such linkages have been demonstrated to occur between myosin and the 11s protein of soy (10,11). Kurth (13) demonstrated the use of transglutaminase to effect cross-linkages between proteins and proposed its application to the binding of restructured meats and to enhance gelation in other muscle food products.

Measurement of Textural Effects Induced by NM Protein Addition to MF Protein Foods

The texture of a gelled muscle food product is primarily the human response during mastication to the mechanical properties of the product, modified to some extent by the moisture release and other mouthfeel-related properties (i.e., greasiness, graininess, etc.) of the gel. The mechanical properties can be fundamentally and quantitatively evaluated in homogeneous, isotropic gels by means of a torsional failure test (14), which effectively measures both the stress and strain required to produce failure (i.e., disrupt the gel). The ratio of stress to strain at failure is the rigidity of the gel at failure, and this has been found to remain approximately constant throughout deformation of most MF protein gels we have tested (i.e., rigidity measured at failure corresponds to low-strain rigidity measurements).

A plot of rigidity vs. strain (Figure 6 (14)) has proven to be quite valuable in evaluating treatment effects in the gelation of MF proteins, in that the movement in textural space can be readily tracked (rigidity indicating gel firmness, and strain indicating gel cohesiveness). From past research of this laboratory, it also seems that the direction of that movement is indicative of the types of changes which are occurring within the protein matrix (14) For example, vertical movement has been found to be very sensitive to protein concentration and free water changes. In Figure 6 (T. Lanier, unpublished data), the greater effect of decreasing moisture content on a

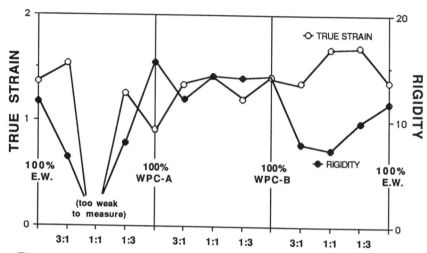

Figure 5   Gel strain and rigidity at failure of egg white (EW), two whey protein concentrates (WPC-A and -B) and their mixtures. Cooked at 90°C for 45 min.

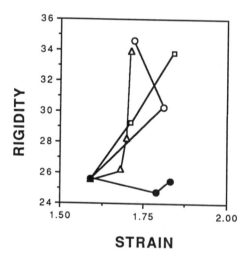

STRAIN

Figure 6   Rigidity-strain plot of heat-set gels prepared from a surimi sol (intercept of all lines) with the following treatments: □ decreasing moisture content; △ dry starch addition; ○ dry whey protein concentrate (WPC) addition; ● WPC addition, holding moisture content identical to the surimi sol.

surimi gel is increased rigidity. Treatments which may induce the formation of greater numbers of bonds, such as addition of bromate to form disulfide bridges, also primarily affect gel rigidity. Perhaps the evenness of the protein dispersion, or "ordering" of the gel matrix is also reflected in vertical movement in such a plot, since low temperature "setting" prior to cooking surimi gels both enhances the gel dispersion (Niwa, E., Ibid.) and primarily affects gel rigidity. For pure MF protein gels, horizontal movement on this diagram has been found to relate closely to the amount of denaturation which has occurred in the proteins prior to gelation, and also to the conditions under which the gel forms (i.e., pH, ionic types and strength). This may be interpreted as indicating that horizontal movement (i.e., gel cohesiveness or compliance) relates to the relative proportion of free amorphous (unbonded) polymer between junction points, and perhaps to the types of bonds (i.e., rigid vs. rotatable) being formed and the continuity of the matrix as well.

Water-binding properties of pure MF protein gels seem also to correlate with horizontal movement on the rigidity strain plot, in that cohesive gels generally have higher water-holding ability as evidenced by timed purge-release tests or expression of moisture by compression through centrifugation or pressing. However, in gels which contain added fillers, this relationship may be lost since the water-holding properties of the filler particles dispersed in the MF protein matrix may overide the water-holding properties that result only from the construction of the matrix itself (15).

When added with no additional water to a formulation, NM proteins increase the protein/solids content and thus effectively reduce the moisture content of the mix. If the NM proteins exert no other influence, the rigidity of the gel will be increased simply due to the reduction in moisture content. Additionally, water is partitioned into the swollen NM protein particles to an extent dependent upon their water-holding capacity, and thereby the concentration of the MF protein in the continuous phase is also increased. Such effects are depicted in Figure 6 for addition of dry starch and whey protein concentrate (WPC) to surimi. At higher levels of dry protein addition, competition between NM and MF proteins for water may ensue, to the extent that gelling properties of the muscle protein may be impaired.

Addition of a dry soy isolate to surimi of lower gelling ability is depicted in Figure 7 (T. Lanier, unpublished data). Again, the effect is primarily to increase the rigidity of the gel, indicating that this particular soy isolate acts mainly as a filler and a water binding agent. Note however that when the soy is added along with water of hydration, the main effect is a decrease in rigidity of the gel. In this latter case, the NM protein is substituting for the MF protein rather than for the water. This constitutes a true test for compatibility and similarity of function between an added NM protein and the MF protein it replaces (16), analogous to the admixtures of egg white and WPC shown in Figure 5. In Figure 6 the addition of WPC to the MF protein gel in substitution for surimi (addition at constant moisture) resulted in increased strain. Such a pronounced deviation from a vertical movement on the rigidity/strain plot due to NM protein addition may suggest that other than strictly filler effects are being exerted by the NM protein.

Figure 8 (E.A. Foegeding; T.C. Lanier, unpublished data) illustrates the effects of substituting a particular soy protein concentrate for water or MF protein in turkey breast or surimi. Simply increasing the protein concentration without changing the protein composition (13:0 vs 10:0) results mainly in increasing the rigidity, very similar to dry addition of starch or protein (Figs. 6, 7). However, substitution of this soy concentrate for MF protein (turkey or surimi; 10:3 vs 13:0) actually increased the cohesiveness (strain) of the gels, with a slight drop in rigidity occurring at the higher MF protein replacement level (7:6 vs 10:3). It is interesting that in these two very different MF protein systems, the trends in the data due to soy addition were identical despite differences in absolute values.

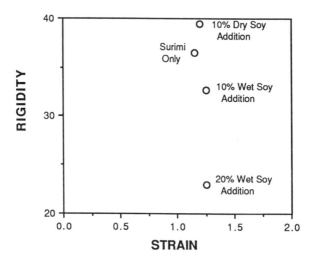

Figure 7   Rigidity-strain plot of heat-set gels prepared from a surimi sol with to which dry or wet (prehydrated to same moisture content as surimi) soy protein isolate was added.

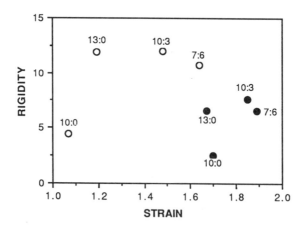

Figure 8   Rigidity-strain plot of heat-set gels prepared from a turkey (O) or surimi (●) sol to which soy protein concentrate was added. Ratios indicate proportion of muscle material: soy in the mixture; sum of the two numbers in the ratio equals the protein content of the mixtures, which varied in moisture content to account for changes in protein content. (E.A. Foegeding; T.C. Lanier, North Carolina State University, unpublished data)

Figure 9 (G. MacDonald, T. Lanier, New Zealand Department of Scientific and Industrial Research internal report, 1989) also illustrates the effects of adding NM protein in substitution for MF protein; in this case a WPC, whole milk protein, and egg white solids are added to a surimi-based model food. In Figure 9a (WPC substitution for surimi) the increase in cohesiveness of the gel (increased strain at failure) at low levels of addition suggests a synergistic interaction with surimi in matrix formation, perhaps by gel entanglement or cooperative linkages. At higher levels of WPC addition however, the effect is a more vertical movement (change in rigidity) with respect to the non-WPC-containing formulation. Figure 9b shows addition of whole milk protein to the same surimi-based gel. Here the gel matrix structure is obviously destabilized by the NM protein addition, for both rigidity and strain decrease. Figure 9c illustrates the addition of egg white solids into the same system. In this case, only the strain is greatly affected, but negatively with respect to the effect of WPC noted in Figure 6 or Figure 9a.

Summary of Factors Differentiating NM Protein Effects in MF Protein Foods

The case studies above serve to illustrate the great diversity of effects that different NM proteins can have on a MF gelling system, even NM proteins from the same source. To summarize, the primary causes of these differences in effect on MF gels can be attributed to the following differences in properties amongst NM proteins:

1. *Effects on the Microenvironment of the Gelling Constituents.* Dry NM proteins differ in water-holding ability, which results in varying levels of competition with MF proteins for water. NM proteins also may possess affinity for certain ions, or they may carry/release ions into the MF protein system (such as calcium) that affect gelation. Additionally, they may affect pH by virtue of the chemistry of their exposed residues and their overall buffering capacity. For example, it has been recently demonstrated (E.A. Foegeding, N.C. State Univ., personal communication) that soy protein concentrates possess considerable buffering capacity such that their addition can adjust the pH of the MF system to a more favorable level for gelation. Similarly, calcium ion addition is known to stimulate the low-temperature "setting" ability of surimi(17), and most WPC preparations carry a considerable concentration of free calcium ions.

2. *Ability to Gel.* NM proteins that do not gel in pure solution would not be expected to gel in combination with MF proteins, and thus more likely function as fillers to the system. Even NM proteins that can gel in pure solution may not be in sufficient concentration, or in the correct microenvironment to gel in a MF protein system. Scanning rheological methods can provide circumstantial evidence of co-gelation of components (see Figures. 2 and 4) but the effects of swelling/water uptake by the NM component on gel rheology at particular temperatures must also be accounted for (Figure 3). If the NM protein does gel, it still may mainly function as a filler by forming discrete particles dispersed in a MF protein gel. It is known that gelling ability of WPC is a highly valued attribute for its inclusion in injection solutions for cured hams in Japan; this may reflect the formation of a gel in discrete pockets within the ham. If the discrete NM protein gel particles also interact with the MF matrix, additional strength or cohesiveness may be obtained. Intimate dispersion and chemical interaction of NM and MF proteins could perhaps lead to establishment of a copolymerized gel network in which gel properties are enhanced; alternatively, the NM protein might instead inhibit cross-reactions of the MF proteins and disrupt gel matrix formation. An interpenetrating network of NM protein in a MF protein matrix formed through intimate dispersion may or may not interact with the MF proteins. Either way, however, the entangling of two gel

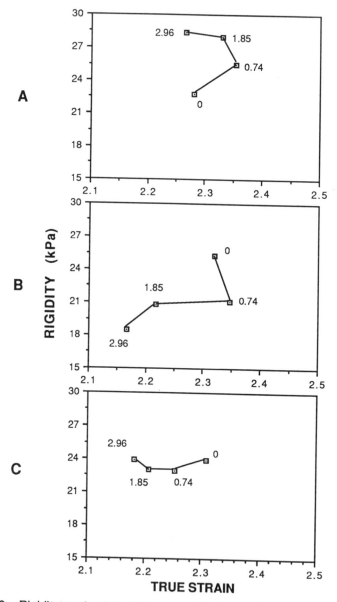

Figure 9   Rigidity-strain plot of heat-set gels prepared from surimi sol to which was added increasing percentages of prehydrated (A) whey protein concentrate (WPC); (B) whole milk protein; and (C) egg white. The moisture contents of all samples were held constant. (G. MacDonald; T.C. Lanier, unpublished data)

networks would be expected to strengthen the gel, or in the case of injected hams, to strengthen the product texture.

3. *Direct Interaction with MF Proteins*. Foegeding and Lanier (1) recently reviewed published reports of direct chemical linkages between NM and MF proteins, most of which were documented in model systems. Gel permeability chromatography, solubility/turbidity measurements, and electrophoresis have provided direct evidence of linkages forming between soy proteins and myosin. However, none of these studies conclusively demonstrated a relationship between chemical linkage of NM to MF protein and enhanced gel strength. One might hypothesize that NM-MF protein linkages would enhance gel strength when the cross-links extend or reinforce the gel matrix, but may have little positive and possibly a negative effect on gel strength when cross-linking leads only to attachment of particles to the matrix (Figure 1C). In the former case, rigid cross-links such as covalent bonds would be expected to affect primarily the gel rigidity and strength (force to failure), as this is the primary effect of covalent cross-linking during "setting" of surimi at low temperatures prior to cooking, and of the enhanced disulfide bonding which is induced by addition of oxidants such as bromate or cystine to surimi sols. Non-rigid bonds, such as hydrophobic interactions and ionic bridges, might allow for enhanced cohesiveness (strain to failure) in the gel.

4. *Filling of Interstitial Spaces in the Gel Matrix*. It has been previously reported that stronger, firmer textures are obtained when non-gelling, and even insoluble, NM proteins are added to muscle food systems than when more functional NM proteins are added (18). These reports thus suggest that water-binding and volume fraction displacement may be the most important roles of added NM proteins in certain processed meat applications. In such a role, NM protein mimics the function of added starch granules which can swell, absorb excess water, and fill the interstitial spaces of the MF protein matrix without interfering with its formation. Indeed, there is evidence that more soluble and functional forms of NM proteins might function to inhibit the gelation properties of the MF protein, much as the undenatured water-soluble protein fraction of surimi has been shown to inhibit gelation of the MF proteins. While Okada and Yamazaki (19) indicated that particle size of an included filler (granule size in the case of added starch) seemed important to the rheological properties of the filled gel, Ziegler (Ziegler, G; Foegeding, E. A. In Advances in Food and Nutrition Research, Academic Press, in press.) demonstrated that it is actually the volume fraction replacement which is more important, as least as regards the elastic modulus (rigidity) of the gel.

Despite evidence that functional NM proteins do not function well as fillers in muscle foods, it may be useful to consider the possibility of employing NM proteins that do gel well to form a "pre-gel" which then can be incorporated as a filler into the MF protein matrix of a food. The functional NM protein could be mixed with added water and certain other constituents of the processed muscle food formulation, heated to form a gel, and then dispersed in the undenatured, comminuted muscle sol prior to heat processing of the food. Such an approach could be particularly useful in replacing lipid globules and crystals (which typically also function as fillers in meat products) to produce low-fat muscle food products. Such factors as the particle size and shape, the volume fraction replacement in the MF system, and the rheology and water-binding properties of the "pre-gel" material would likely be important to the effects its inclusion might effect on properties of the finished product.

A Proposed Approach to the Study of NM Protein Interactions in MF Protein Foods

It is appropriate that much of the literature to date documenting NM-MF protein gelling interactions has been carried out in model systems, since they provide the controlled conditions necessary for documenting the contribution of each component in the system. However, it becomes necessary subsequently to extend these studies to more complex systems that approximate or recreate the processing conditions of the food manufacturing plant, so that fundamental concepts can be applied in practice. This would hopefully eventually eliminate the trial-and-error approach now practiced by food ingredient and food product manufacturers in the selection of ingredients for particular applications. Figure 10 presents a suggested approach to the investigative process that should be undertaken to document the physico-chemical basis of NM protein functionality in gelled muscle foods.

Initially it is important to understand how the composition and processing/storage/handling of the NM protein affect its physico-chemical properties. What factors affect its surface chemistry, and how does this relate to its functionality (solubility, water-holding, and gelation ability) in the "pure" state? How can this be manipulated by changes in the processing environment? Next, it must be determined how this functionality in the "pure" state, and the surface chemistry responsible, relates to the manner in which the NM protein affects a simple MF protein gel. Perhaps the NM protein can be manipulated to perform either as a filler or a co-gelling ingredient dependent upon the processing conditions employed. Conditions can be identified at which the NM protein acts synergistically with MF protein to build a strong, cohesive gel, or antagonistically to yield a more brittle or soft structure. Either result may be desired, depending upon the application.

Once an understanding is gained of the surface chemistry and functionality of the NM protein, both alone and incorporated into a MF protein gel, the investigation should proceed toward a specific food application. The first step would be to test the ingredient in a model food system, this being compositionally similar to the commercial product of interest but prepared under more controlled laboratory conditions on bench-scale equipment. Deviations from the performance of the NM protein in this system versus the plain MF protein system may be due to interactions of the proteins with the added non-protein components of the model food. If no new interactions between NM protein and the MF protein-based model food are discovered (or if they are found to be predictable), then either this model food system or the plain NM-MF protein gel system may be useful as a routine quality test for NM protein to be used in the food system. This assumes, of course, that when the NM protein is finally tested in the actual food formulation under manufacturing conditions (the last step in this investigative process), that no startling deviations from prior performance in more model systems are discovered.

It is particularly important, in the course of such studies, that the rheological properties of the model systems be measured in fundamental terms. Too often a report employing excellent biochemical measurements is flawed by correlation only with empirical textural tests, and the underlying structure-function relationship is obscured. There is no reason why physico-chemical measurements may not some day be used to predict functionality in a complex food system if all the measurements are made in the most fundamental terms possible. Li-Chan et al. (20) proposed that such measurements can be used as coefficients for ingredient variables in linear programming of food formulations in order to select ingredients which, when intermixed and appropriately processed, will produce the desired end-product attributes. The ultimate end of studies involving the functionality of NM proteins in muscle foods is, after all, to be able to predict which proteins (and why) will function best in each application. Linear programming is now commonly used in a crude manner for cost and composition control of many muscle foods. Systematic study of

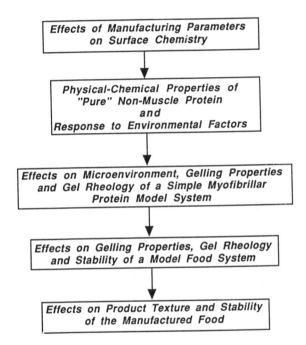

Figure 10 Proposed approach to the study of non-muscle protein effects in myofibrillar protein foods.

NM-MF interactions will lead to more precise cost and quality control of muscle foods in the future.

Conclusion

In the large majority of cases, it is likely that NM proteins added to a processed meat product function chiefly as fillers of a MF protein matrix. However, even in this role, they may indirectly affect gelation of the MF proteins by altering the biochemical environment and thus alter the nature of the continuous matrix in which they and other food constituents are entrapped. Additionally, the texture of the food is influenced as much by the volume fraction and nature of entrapped constituents as it is by the properties of the continuous matrix. Thus, whether NM proteins act as fillers or gelling adjuncts to the MF proteins, their impact on meat product texture can be considerable.

Literature Cited

1. Foegeding, E. A.; Lanier, T.C. Cereal Foods World 1987, 32, 202.
2. Wicker, L.; Lanier, T.C.; Knopp, J.A.; Hamann, D.D. J. Agr. Food Chem. 1989, 37, 18.
3. Lanier, T.C. Food Technol. 1986, March, p107.
4. Wu, M.C.; Lanier, T.C.; Hamann, D.D. J. Food Sci. 1985, 50, 20.
5. Tolstoguzov, V.B. In Functional Properties of Food Macromolecules; Mitchell, J.R.; Ledward, D.A., Ed.; Elsevier: New York, 1986; p 385.
6. Burgarella, J.C.; Lanier, T.C.; Hamann, D.D. J. Food Sci. 1985, 50, 1588.
7. Nakagawa, T.; Nagayama, F.; Ozaki, H.; Watabe, S.; Hashimoto, K. Nippon Suisan Gakkaishi 1989, 55, 1045.
8. Shimizu, Y.; Nishioka, F. Nippon Suisan Gakkaishi 1974, 40, 231
9. Clark, A.H; Richardson, R.K.; Robinson, G.; Ross-Murphy, S.B.; Weaver, A.C. Prog. Fd. Nutr. Sci. 1982, 6, 149.
10. Haga, S.; Ohashi, T.; Yamauchi, K. Jap. J. Dairy Food Sci. 1984, 33, 131.
11. Peng, I.C.; Dayton, W.R.; Quass, D.W.; Allen, C.E. J. Food Sci. 1982, 47, 1984.
12. Foegeding, E.A.; Dayton, W.R.; Allen, C.E. J. Food Sci. 1986, 51, 109.
13. Kurth, L. Food Technol. Austral. 1983, 35, 420.
14. Hamann, D.D.; Lanier, T.C. In Seafood Quality Determination; Kramer, D.E.; Liston, J., Ed.; Elsevier: New York, 1987, p 123.
15. Foegeding, E.A.; Ramsey-Bottcher J. Food Sci. 1987, 52, 549.
16. Iso, N; Mizuno, H.; Saito, T; Lin, C.Y.; Fujita, T.; Nagahisa, E. Nippon Suisan Gakkaishi 1985, 51, 485.
17. Saeki, H; Wakameda, A.; Ichihara, Y.; Sasamoto, Y. Nippon Suisan Gakkaishi 1989, 55, 1867.
18. Yasumatsu, L.; Misaki, M.; Tawada, T.; Sawada, K.; Toda, J.; Ishii, K. Agric. Biol. Chem. 1972, 36, 737.
19. Okada, M.; Yamazaki, A. Nippon Suisan Gakkaishi 1959, 25, 440.
20. Li-Chan, E.; Nakai, S.; Wood, D. F. J. Food Sci. 1987, 52, 31.
21. Zieglar; Fageding. The Gelation of Proteins: Advances in Food and Nutrition Research; Academic Press: Orlando, Fla., in press.

RECEIVED June 20, 1990

Author Index

Aoki, Takayoshi, 164
Arntfield, S. D., 91
Arteaga, G., 42
Barford, Robert, ix
Brekke, Clark J., 257
Brown, E. M., 182
Closs, B., 137
Colas, B., 137
Cooke, P., 25
Cornell, Donald G., 122
Courthaudon, J. L., 137
Creamer, Lawrence K., 148
Damodaran, Srinivasan, 104
Farrell, H. M., Jr., 182
Fligner, Karen L., 1
Foegeding, E. Allen, 257
Hamann, D. D., 212
Haylock, Steve J., 59
Hirotsuka, M., 42
Ismond, M. A. H., 91
Kakalis, L. T., 182

Kato, Akio, 13
Kumosinski, T. F., 182
Lanier, Tyre C., 268
Le Meste, M., 137
Li-Chan, E., 42
Loh, Jimbay, 228
Mangino, Michael E., 1
Matsudomi, Naotoshi, 73
Murray, E. D., 91
Nakai, S., 42
Oh, Sangsuk, 195
Parris, N., ix,25
Pessen, H., 182
Richardson, Tom, 195
Sanderson, Wayne B., 59
Smith, Denise M., 243
Song, Kyung B., 104
Vazquez, M. C., 42
Woychik, J. H., 25
Xiong, Youling L., 257

Affiliation Index

Ecole Nationale Superieure de Biologie
    Appliquee a la Nutrition et a
    l'Alimentation, 137
Kagoshima University, 164
Kraft General Foods, 228
Michigan State University, 243
New Zealand Dairy Research Institute,
    59,148
North Carolina State University,
    212,257,268

Ohio State University, 1
U.S. Department of Agriculture,
    25,122,182
University of British Columbia, 42
University of California, Davis, 195
University of Kentucky, 257
University of Manitoba, 91
University of Wisconsin–Madison, 104
Washington State University, 257
Yamaguchi University, 13,73

Subject Index

A

Adsorption mechanisms, effect on
emulsification, 6
Adsorption of proteins at air–water interface
adsorption isotherm of bovine serum
albumin intermediates, 112,113f,114
adsorption study procedure, 106–107
bovine serum albumin
structural intermediates
conformational characteristics, 107,109t
preparation, 106
bovine serum albumin intermediates bulk
phase concentration vs. constant plots, 119f
surface concentration vs. half-
time plots, 109,110–111f
surface concentration vs. time plots,
117,118f,119
surface pressure isotherm, 112,113f,114
β-casein, surface concentration
vs. half-time plots, 114,115f
circular dichroism
bovine serum albumin intermediates,
spectra, 107,108f
procedure, 106
conformational characteristics of bovine serum
albumin structural intermediates, 107,109t
diffusion coefficients of bovine serum
albumin and β-casein, 109,112t
experimental materials, 105–106
hydrodynamic radius determination, 106
phenomenological rate law, 117
reversibility or irreversibility, effects, 117
surface energy barrier, effect, 116
Adsorption of proteins at liquid interfaces
dynamics, 104–105
role of protein conformation, 105
Aggregation between lysozyme and
heat-denatured ovalbumin
degree of heat denaturation, effects, 82,84f
elution patterns of supernatants, 82,85f,86
NaCl concentration, effects, 82,84f
weight ratio of lysozyme to
ovalbumin, effects, 80,82,83f
Amino acid hydrophobicity constants
correlation with peptide chromatographic
retention times, 47,49t,50
limitations, 50

Axial compression, property of food gels
related to texture, 218–220

B

Basic–acidic protein mixtures, effect on
foaming, 9
Bovine casein, quaternary structural changes
by small-angle X-ray scattering, 183–193
Bovine κ-casein, genetic engineering to
enhance proteolysis by chymosin, 196–210
Bovine casein micelles
aggregates cross-linked by colloidal
calcium phosphate
dissociation during dialysis, 168–172
dissociation of individual casein
constituents, 171,172f
heating and cooling effects, 179
incorporation of individual casein
constituents, 174–175,176–177f
partition of individual casein
constituents, 175,177f
Ca content, 167t
calcium phosphate, colloidal,
See Colloidal calcium phosphate
casein aggregates cross-linked by colloidal
calcium phosphate and inorganic phosphate,
changes in content, 175,176f
casein composition, 168t
casein composition of dialyzed micelles,
168–169f,171t
colloidal calcium phosphate
casein interaction sites, 178t,179
cross-linkage
heating and cooling effects, 179
individual milk salt constituents,
role, 173t,174
composition, 164
disaggregated micelles, elution
patterns, 165,166f,167
disaggregation, 164,168
elution patterns during dialysis, 168,170f,171
inorganic P content, 167t
micellar casein content, changes
during dialysis, 168,169f
separation, 165,166f,167
structural model, 164

Bovine serum albumin, adsorption at air–water interface, 105–119

C

Calcium, effect on gel formation, 2
Calcium ions, effect of phospholipid–whey protein interactions, 132–133,134f,135
Carbon-13 NMR spectroscopy, molecular dynamics of casein, 191,192f,193
Casein(s)
description, 26
existence as micelles in milk, 148
α-Casein, presence in milk, 149–161
$α_{s1}$-Casein, characteristics, 154
$α_{s2}$-Casein, characteristics, 154
β-Casein
adsorption at air–water interface, 105–119
characteristics, 154
film formation, 7
presence in milk, 149–161
κ-Casein
accessibility, effect on casein micellar structure, 153
bovine, See Bovine κ-casein
characteristics, 154–155
function, 210
hydrolysis by chymosin, 196,198f
presence in milk, 149–161
stabilization of casein micelle, 151–152
Casein micellar dispersion, preparation, 168
Casein micelle(s)
bovine, See Bovine casein micelles
composition, 149t
definition, 182
description, 92
formation, 149
models, 149–150,182
molecular dynamics by ¹³C-NMR spectroscopy, 191,192f,193
quaternary structural changes by small-angle X-ray scattering, 184–191
solvation, 155–156
structure
accessibility of κ-casein, 153
circular dichroism spectra, 158–159
colloidal calcium phosphate, role, 152–153
electron microscopic data, 154
mammalian milk, 156,157t,158

Casein micelle(s)
structure—Continued
micelles with different ratios of casein components, 156
micelles with divalent cations and without colloidal phosphate, 155
model(s)
proposed, 182
selection criteria, 150–159
NMR studies, 159
problems, 159–161
rate of divergence of casein sequences, 158
siting of κ-casein, 159–160
small-angle neutron scattering studies, 159
small-angle X-ray scattering studies, 159
specific side-chain interactions, 160–161
stabilization of micelles,
nature of forces, 150–152
submicelles, See Casein submicelles
water content, 155–156
Casein proteins, 60t,61,62t
Casein submicelles
molecular dynamics by ¹³C-NMR spectroscopy, 191,192f,193
quaternary structural changes by small-angle X-ray scattering, 184–191
Cheese
bond hydrolysis rates, alteration by genetic engineering, 196–210
formation, role of bond hydrolysis, 195–196
genetic engineering techniques, alteration of bond hydrolysis rates, 196–210
Chymosin
cheese curd formation, role, 195
hydrolysis of κ-casein, 196,198f
partial purification and hydrolysis of expressed κ-casein, 205,207–210
Colloidal calcium phosphate
bovine casein micelle component, 164
casein interaction sites, 178t,179
casein micellar structure, role, 152–153
cross-linkage in casein micelles, effect of heating and cooling, 179
incorporation of individual casein constituents into cross-linked casein aggregates, 174–175,176–177f
milk salt constituents, role in cross-linking, 173,174t
Complex gel models, partitioning of nonmuscle protein in myofibrillar protein food, 272,274f

Complex modulus, 215–216
Conalbumin, rheological study of gelation, 216,217f
Conformation of proteins, effect on adsorption at air–water interface, 105–119
Constant stress rheometry of food dough
  bread crumb, force vs. deformation curve, 232,235f
  bread crumb texture, 237t,240,241t
  bread preparation, 232
  bread sampling procedure, 232,233f
  bread testing procedure, 232,233f,234,235f
  creep parameter(s), 234,236t,237,238–239f,240,241t
  creep parameters of dough vs. textural quality of bread crumbs, 240t
  crumb extensibility, 232,235f
  crumb strength, 232,235f
  data analysis, 234
  dough preparation, 231,233f
  dough testing procedure, 231–232
  loaf density, effect of degree of mixing, 240,241t
  sensory toughness vs. instrumental toughness, 234,235f
Cow milk, casein compositions, 156,157t,158
Creep, 229
Crumb extensibility, 232,235f
Crumb strength, 232,235f

D

Denaturation of whey protein, effect on emulsifying properties of whey proteins, 26
Disulfide bonds, effect on surface hydrophobicity, 7
Dough, handling properties and texture, effect of consistency and rheology, 228
Dynamic oscillation method, limitations for food dough rheological analysis, 228

E

Egg replacer
  development, 68
  performance comparison with that of whole egg, 69–70t
Egg white, importance of heat coagulability, 73

Electrolytes, effect on gel formation, 2
Electron spin resonance (ESR) spectroscopy
  description, 138
  milk protein–lipid interactions, 137–138
  nitroxide radicals, 138,140f
Electrostatic repulsion, role in emulsifying properties of food proteins, 15,17,18f
Emulsification, 4–8
Emulsification activity index
  definition, 4
  influencing factors, 5–8
  nonfat dry milk powders, 36,39f
Emulsifying activity
  correlation with hydrophobicity, 44
  definition, 4
  influencing factors, 5–8
Emulsifying capacity
  definition, 4
  influencing factors, 5–8
Emulsifying properties of food proteins, role of electrostatic repulsion, 15,17,18f
Emulsion(s), 4
Emulsion stability
  definition, 4
  influencing factors, 5–8
Enzymatic modifications, effect on foaming, 9–10
Extracted protein concentration, effect on gelation of myofibrillar protein, 264

F

Filled gel models, partitioning of nonmuscle protein in myofibrillar protein food, 270,271f,272
Fillers, 250
Fluorescence probe methods, measurement of protein hydrophobicity, 50–51
Foam formation, nonfat dry milk powders, 39,40f
Foam stability, nonfat dry milk powders, 39,40f
Foaming, 8–10
Foaming properties of food proteins, protein–protein interaction, 15
Food dough, constant stress rheology, 231–241
Food gels, fundamental properties related to texture, 218–226

Food protein(s)
  bond hydrolysis rates, effects on
    functional characteristics, 195
  macromolecular interaction and stability,
    effect on functional properties, 13–23
Food protein gels, 213
Food systems
  interactions, 65–70
  milk protein ingredients, 62,63t
Fracture shear strain, 219
Functional properties of food proteins
  correlation coefficients of Gibbs energy
    with surface properties, 19,23t
  electrostatic repulsion, role in
    emulsifying properties, 15,17,18f
  experimental procedure, 14–15
  foaming properties
    mutants of substituted tryptophan
      synthase, 19,20f
    protein–protein interactions, effects, 15,16f
  Gibbs energy of unfolding, 17,19,20t
  Gibbs energy vs. emulsifying activity, 19,22f
  Gibbs energy vs. foam stability, 19,22f
  Gibbs energy vs. foaming power, 19,21f
  Gibbs energy vs. surface tension, 19,21f
  protein–protein interactions, effect on
    foaming properties, 15,16f
  protein stability, significance, 17–23
  research, 13
  structural properties, relationship, 23t
  surface hydrophobicity of proteins, 13–14

G

Gapped heteroduplex method, advantages, 203
Gel formation, requirements, 1–2
Gel fracture, 218
Gel matrix formation by myofibrillar
  proteins, 268–270
Gelation, 1–3
Gelation of myofibrillar protein
  activation energy vs. protein source, 261,263t
  extracted protein concentration, effects, 264
  hydrophobicity vs. thermal input, 261,262f
  myosin–actomyosin ratio, 260t,261
  pH effects, 264,265f,266
  protein association vs. thermal
    transition, 261,262f,263t,265f
  protein concentration, effects, 258,259f,t,260

Gelation of myofibrillar protein—Continued
  turbidity change vs. protein source,
    261,262f,263,265f
Gelation of proteins
  food formation, 212
  rheology, 212–226
Genetic engineering of bovine κ-casein
  bacterial strains and plasmids, 196
  κ-casein, expressed
    detection in Escherichia coli, 201
    partial purification and hydrolysis
      by chymosin, 205,207–210
    quantitation, 205
    κ-casein, role, 210
  chymosin hydrolysis patterns for mutant and
    wild-type κ-caseins, 205,208f
  DE–52 anion-exchange chromatographic
    and dot blot analyses of fractions, 205,207f
  DNA sequence of 5′ region, 203,204f
  expression vector, 200,203,205,206f
  hydrolysis rates of wild-type and mutant
    κ-casein, 205,209f,210
  mutagenic reaction procedure, 197
  mutant screening procedure, 197,200
  oligonucleotide-directed mutagenesis,
    197,199f,202–203,204f
  oligonucleotide sequence, 197,198f
  oligonucleotide synthesis and
    purification, 202
  partial purification of κ-casein, 201
  single-stranded template preparation, 197
  template preparation analysis, 202
Goat milk, casein composition, 156,157t,158

H

Heat-induced aggregation
  egg white protein, role of lysozyme, 73–89
  ovalbumin and lysozyme
    aggregation scheme, 80,81f
    carboxymethylation of sulfhydryl
      groups, 80,81f
    ionic strength and heating temperature,
      effects, 77,79f
    sodium dodecyl sulfate electrophoresis
      and densitometric scans, 77,78f
    succinylation of amino groups,
      77,79f,80
  ovotransferrin and lysozyme, 86,87f

Heat-induced gelation of muscle proteins
  experimental materials, 244
  functionality of salt-soluble protein
    fractions, 245,246t
  gel evaluation procedure, 245
  gel preparation procedure, 244–245
  importance, 243–244
  salt-soluble protein fractions, composition
    effects, 246,247t
  salt-soluble protein gel properties
    nonmeat proteins, effect, 250,252–254t
    pH effects, 248,249f,250,251f
    stromal and sarcoplasmic proteins,
      effects, 246–247,248t
    temperature effects, 248,250,251f
  solubility differences of skeletal,
    cardiac, and smooth muscle proteins, 245t
Heat-set gel rheology, effects of muscle and
  nonmuscle protein interactions, 268–284
Heat stability of lysozyme-free egg white
  protein
  aggregation, effect of heating temperature
    and added lysozyme, 75,76f
  sodium dodecyl sulfate–polyacrylamide
    gel electrophoretic patterns, 75,76f
Hencky's strain, 219
Human milk, casein compositions,
  156,157t,158
Hydrophobic properties of milk proteins,
  effect of pre-heat temperature, 26–41
Hydrophobicity
  determination, 51–56
  emulsification, effects, 6
  foaming, effects, 9
  food protein functionality, role, 43
  gel formation, effects, 3
  indicators, 43
  laser Raman spectroscopic determination,
    54,55f,56t
  measurement methods, comparison, 44,47–51
  proton NMR determination, 51,52–53f,54
  structure–activity relationships of
    food proteins, 44–54

I

Interactions in food systems
  egg replacer, 65,68–70t
  ham products, 63–67

Ionic strength, effect on foaming, 9
Ionic type and strength, effect on gel
  matrix formation, 269

L

Laser Raman spectroscopy, determination of
  protein hydrophobicity, 54,55f,56t
Linear fracture mechanics, description, 212
Linear programming, use in food
  formulation, 54,57
Lipid–milk protein interactions, See Milk
  protein–lipid interactions
Loss tangent, 216
Lysozyme, film formation, 7
Lysozyme, role in heat-induced aggregation
  of egg white protein
  chemical modification of proteins, 74
  gel electrophoretic procedure, 75
  gel filtration procedure, 75
  gel hardness, measurement, 74
  lysozyme and heat-denatured ovalbumin,
    aggregation, 80,82–86
  lysozyme-free egg white protein,
    heat stability, 75,76f
  ovalbumin, heat-induced gelation, 86,88f,89
  ovalbumin and lysozyme, aggregation, 77–81
  ovotransferrin and lysozyme,
    aggregation, 86,87f
  protein preparation, 74
  turbidity, measurement, 74

M

Micellar aggregates, formation, 91
Micelle, definition, 91
Milk protein(s)
  composition, 60,61t
  emulsion stabilization mechanism,
    factors affecting, 137
  foaming, effects, 8–9
  functional properties, 61,62t
  hydrophobic properties, effect of
    pre-heat temperature, 26–41
  isolation, 60
  properties, 60t
  protein modification techniques, 61,62t
  role in food systems, 61

Milk protein ingredients, use in food
    systems, 59,62,63t
Milk protein–lipid interactions
  amino residues of milk proteins,
    modifications, 144,146,147f
  disulfide bridge reduction and
    sulfhydryl carboxymethylation
    procedures, 138
  effects, 146
  emulsion preparation, 139
  ESR analytical procedure, 139
  ESR spectra, 139,140f
  fatty acids incorporated into triglyceride
    droplets, behavior, 142,143t
  glycosylation procedure, casein, 138
  lysine residues of caseins, effect of
    blockage by glycosylation, 146
  mobility
    apolar chain, 142,143t
    polar end of fatty acid, 141t,142
  protein sample preparation, 138
  protein side chains, motion, 144t,145f
  trypsic hydrolysis of b-lactoglobulin
    effects, 146,147f
    procedure, 138
Mobility, parameters, 139
Molecular flexibility, effect on foaming, 10
Muscle protein(s), factors affecting
  heat-induced gelation, 243–254
Muscle protein gels, factors affecting
  texture and water holding, 258
Mutant DNA, screening procedures, 203
Myofibrillar protein(s), factors affecting
  gel matrix formation, 268–270
Myofibrillar protein gelation, See
  Gelation of myofibrillar protein
Myosin, role in sausage structure, 257–258
Myosin–actomyosin ratio, effect on
  gelation of myofibrillar protein, 260t,261

N

Nonfat dry milk
  applications, 26
  bread crumb texture, effects, 237t,240
  creep parameters of food dough,
    effects, 234,236t,237,238–239f
  emulsifying activity index, 36,39f
  foam formation, 39,40f

Nonfat dry milk—Continued
  foam stability, 39,40f
  functional properties, effect of heat
    treatment, 26
  hydrophobic properties, effect of
    pre-heat treatment, 26–41
Nonmeat proteins, effect on salt-soluble
  protein gel properties, 250,252–254t
Nonmuscle protein(s)
  gelation ability, 279,281
  gelling constituents, effect on
    microenvironment, 279
  interstitial spaces in gel matrix, filling, 281
  myofibrillar protein interactions, 281
Nonmuscle protein interactions in
  myofibrillar protein foods, proposed
  study approach, 282,283f,284
Nonmuscle protein partitioning in
  myofibrillar protein food
  measurement of textural effects, 275–280
  models, 270–275

P

Partitioning of nonmuscle protein in
  myofibrillar protein food
  gel strain and rigidity at failure of
    gelation, 272,275,276f
  models, 270–275
    complex gel models, 272,274f
    filled gel models, 270,271f,272
    thermal scanning plots, 272,273–274f
Peptide chromatographic retention times,
  correlation with amino acid
  hydrophobicity constants, 47,49t,50
pH effects
  emulsification, 5
  foaming, 9
  gel formation, 3,269
  myofibrillar protein gelation, 264,265f,266
  phospholipid–whey protein interactions
    129,131t,132
  protein micelles, 94,95f,96
  salt-soluble protein gel properties,
    248,249f,250,251f
Phospholipid(s), stabilization of milk
  lipid phase, 122–123
Phospholipid–protein interactions,
  influencing factors, 123

Phospholipid–whey protein interactions
  calcium ions, effects, 132–133,134f,135
  circular dichroism spectra of
    monolayers, 132,134f
  experimental procedure, 123–124,126–127
  protein binding to monolayer lipids
    determination, 129,
  pH effects, 129,131t,132
  surface pressure of phospholipid, increase
    initial film pressure, comparison, 129,130f
  injection of protein into subphase,
    127,128f,129
  surface pressure of proteins vs.
    concentration, 127,128f
pI, effect on emulsification, 5–6
Plaque hybridization, use for mutant DNA
  screening, 203
Polarity, effect on foaming, 9
Pre-heat temperature, effect on hydrophobic
  properties of milk proteins
  antiserum to cow whey protein against
    milk, Western blot, 34,35f,36
  electron microscopic procedure, 28
  electrophoretic blotting procedure, 27
  experimental materials, 27
  functional properties, 29,36,39–40f
  gel electrophoretic procedure, 27
  heptane binding, 34t
  HPLC separations, 29–34
  hydrophobic interaction,
    chromatographic procedure, 27
    chromatographic profiles, 29,32–33f,34
  hydrophobicity, determination, 28
  immunogold-labeled rehydrated nonfat
    dry milk powders, electron
    micrographs, 36,38f
  immunogold-labeled skim milk,
    electron micrographs, 36,37f
  immunological detection procedure,
    27–28
  reversed-phase HPLC elution
    profile, 29,30–31f
  Western blot, antiserum to cow whey
    protein against milk, 34,35f,36
  whey protein complexes, location, 34–38
Preparation technique, effect on gel
  formation, 3
Processed meats, variation in quality, 257
Processing techniques, effect on
  emulsification, 5

Protein(s)
  emulsions, stabilization, 4
  stability to enzymatic attack, 195
Protein adsorption at interfaces,
  influencing factors, 4
Protein composition, effect on gel formation, 2
Protein concentration, effect on gelation
  of myofibrillar protein gels,
    258,259f,t,260
Protein conformation, effect on adsorption
  at the air–water interface, 105–119
Protein content, effect on gel formation, 2
Protein gelation, See Gelation of proteins
Protein hydrophobicity
  gel formation, effects, 3
  proton NMR spectroscopy, determination,
    51,52–53f,54
  See also Surface hydrophobicity of proteins
Protein micellar mass, 92
Protein micelle(s)
  applications for networks in food
    systems, 102
  examples, 92
  interaction, 92,93f,94
  noncovalent forces in formation, 101–102
  pH effects, 94,95f,96
  photographs of micelles from various
    sources, 92,93f
  response rating, 94
  salt effects, 96,97f,98
  sucrose effects, 98,99f
  urea effects, 98,100f,101
Protein modifications, effect on
  emulsification, 7
Protein–protein interactions, effect on
  foaming properties of food proteins, 15,16f
Protein stability, significance in functional
  properties of food proteins, 17–23
Proton NMR spectroscopy, determination
  of protein hydrophobicity, 51,52–53f,54
Pure shear, 220

Q

Quantitative structure–activity
  relationship (QSAR) of food proteins
  descriptor selection, 43–44,45t,46f
  formula optimization, applications, 54,57
  physicochemical descriptors, 44,45t

Quantitative structure–activity relationship (QSAR) of food proteins—*Continued*
  protein concentration, effect on emulsifying activity, 44,46*f*
  protein hydrophobicity measurement, 44–54
Quantitative structure–activity relationship (QSAR) techniques, 42
Quaternary structural changes of bovine casein by small-angle X-ray scattering
  distance distribution of micelles, 188,189–190*f*,191
  hydration dynamics and geometry of casein-bound water
    micelles, 187*t*,188
    submicelles, 185,187*t*
  micelles, 184–191
  molecular modeling procedure, 183
  molecular parameters for variants A and B
    micelles, 184*t*,186–187
    submicelles, 184*t*,185
  NMR measurement procedure, 183
  sample preparation, 183
  small-angle X-ray scattering measurement and data analytical procedures, 183
  structural parameters for variants A and B
    micelles, 185*t*,187
    submicelles, 185*t*,186
  submicelles, 183–187

R

Restriction endonuclease digestions, use for mutant DNA screening, 203
Reverse micelles, 91
Rheology, 212
Rheology of food dough
  characterization techniques, 228–229
  fundamental measurements, 228
  models for behavior, 229,230*f*
  stress-softening phenomena, 231

S

Salt effects
  bread crumb texture, 237*t*,240
  creep parameters of food dough, 236*t*,237,238–239*f*

Salt effects—*Continued*
  gel formation, 3
  protein micelles, 96,97*f*,98
Salt-soluble protein, effect of isolation method on protein content, 258
Salt-soluble protein suspensions, composition, 258
Sarcoplasmic proteins, effect on salt-soluble protein gel properties, 246–247,248*t*
Sensory toughness, relationship to instrumental toughness, 232,235*f*
Shear, 213
Shear modulus, 213
Shear rate, 215
Shear strain
  definition, 213
  protease inhibitors and potato starch, effects, 223,225*f*,226
  torsion testing, applications, 223,224*f*
Shear stress
  definition, 213,215,219
  protease inhibitors and potato starch, effects, 223,225*f*,226
  torsion testing, applications, 223,224*f*
Sheep milk, casein compositions, 156,157*t*,158
Shortening effects
  bread crumb texture, 237*t*,240
  creep parameters of food dough, 234,236*t*,237,238–239*f*
Sinusoidal shear strain, 215
Small-angle X-ray scattering
  bovine casein, analysis of quaternary structural changes, 183–193
  description, 182–183
Sol–gel transition rheology
  axial compression, 218–220
  conalbumin, 216,217*f*
  rheological data, physical meaning, 215–216
  rheometer, 213,214*f*
  shear deformation, representation, 213,214*f*
  shear stress and shear strain data, applications, 223,224–225*f*,226
  test modes, 213,215
  theory, 213
  torsion testing of protein gels, 220–224
Solubility, use as QSAR descriptor, 43
Stress relaxation, use for food dough rheological analysis, 228–229

Stromal proteins, effect on salt-soluble
  protein gel properties, 246–247,248t
Structural properties of food proteins,
  relationship to functional
  properties, 23t
Submicelles, 182,183–187
  See also Casein submicelles
Sucrose effects
  bread crumb texture, 237t,240
  creep parameters of food dough,
    234,236t,237,238–239f
  protein micelles, 98,99f
Sulfhydryl(s), effect on gel formation, 2–3
Sulfhydryl groups, effect on
  emulsification, 6
Surface hydrophobicity of proteins, effect
  on functional properties of food
  proteins, 13–14

T

Temperature effects
  gel matrix formation, 269
  salt-soluble protein gel properties,
    248,250,251f
Textural effects induced by nonmuscle
  protein addition to myofibrillar
  protein foods
  moisture effects, 277,278f
  nonmuscle protein substitution for
    myofibrillar protein, effects, 279,280f
  soy protein concentration, effects, 277,278f
  treatment effects, 275,276f,277
Thermal scanning rheology
  limitation, 218
  protein gelation analysis,
    216,217f,222–223,224f

Thermally induced protein gelation, use of
  rheology for study, 212–226
Time, effect on gel matrix formation, 269
Torsion testing of protein gels,
  220,221f,222–223,224f
Triglycerides, description, 143

U

Urea, effect on protein micelles, 98,100f,101

V

Viscosity, 213

W

Water, effect on gel matrix formation, 268–269
Water-added ham products, 63
  manufacturing processes, 64t
  product characteristics, 64,65t
  response surface diagrams, 65,66–67f
Whey protein(s)
  composition, 60,61t
  denaturation, 26
  functional properties, 61,62t
  isolation, 60
  properties, 60t
Whey protein concentrate gels, factors
  affecting formation, 2–4
Whey protein isolate gels, factors
  affecting formation, 3
Whey protein–phospholipid interactions,
  See Phospholipid–whey protein
  interactions

Production: Victoria L. Contie
Indexing: Deborah H. Steiner
Acquisition: Barbara C. Tansill

Books printed and bound by Maple Press, York, PA

Paper meets minimum requirements of American National Standard
for Information Sciences—Permanence of Paper for Printed Library
Materials, ANSI Z39.48–1984   ∞